Intelligent Systems, Control and Automation: Science and Engineering

Volume 85

More information about this series at http://www.springer.com/series/6259

Gilberto Reynoso Meza · Xavier Blasco Ferragud
Javier Sanchis Saez · Juan Manuel Herrero Durá

Controller Tuning with Evolutionary Multiobjective Optimization

A Holistic Multiobjective Optimization Design Procedure

Springer

Gilberto Reynoso Meza
Pontifícia Universidade Católica do Paraná
Curitiba, Paraná
Brazil

Javier Sanchis Saez
Universitat Politècnica de València
Valencia
Spain

Xavier Blasco Ferragud
Universitat Politècnica de València
Valencia
Spain

Juan Manuel Herrero Durá
Universitat Politècnica de València
Valencia
Spain

ISSN 2213-8986 ISSN 2213-8994 (electronic)
Intelligent Systems, Control and Automation: Science and Engineering
ISBN 978-3-319-41299-3 ISBN 978-3-319-41301-3 (eBook)
DOI 10.1007/978-3-319-41301-3

Library of Congress Control Number: 2016943806

Printed on acid-free paper

This Springer imprint is published by Springer Nature
The registered company is Springer International Publishing AG
The registered company address is: Gewerbestrasse 11, 6330 Cham, Switzerland

Preface

In this book we summarise the efforts and experiences gained by working with multiobjective optimization techniques in the control engineering field.

Our studies began with an incursion into evolutionary optimization and two major control systems applications: controller tuning and system identification. It quickly became evident that using evolutionary optimization in order to adjust a given controller is helpful when dealing with a complex cost function. Nevertheless two issues were detected regarding the cost function: (1) sometimes minimizing a given index fails to guarantee the expected performance (that is, when it is implemented); and (2) it fails to reflect properly the expected trade-off between conflictive design objectives. The former issue is sometimes simply that the index does not accurately reflect what the control engineer really wants, the latter because sometimes it is difficult to built a cost index, merging all design objectives and seeking a desired balance among them.

That is how multiobjective evolutionary optimization entered into the scene. Sometimes when aggregating design objectives in order to create a single index for optimization, some understanding of the outcome solution is lost. With multiobjective optimization it is possible to work with each design objective individually. Furthermore it is possible to analyse, at the end of the optimization process, a set of solutions with a different trade off (the so called Pareto front). Therefore, it is possible to select a given solution, with the desired balance between conflictive design objectives.

From there, a lot of work has been carried out on identifying applications, developing optimization algorithms and developing visualization tools. The book is part of a bigger research line in evolutionary multiobjective optimization techniques. Its contents focus mainly on controller tuning applications; nevertheless, its ideas, tools and guidelines could be used in different engineering fields.

Curitiba, Brazil
Valencia, Spain
April 2016

Gilberto Reynoso Meza
Xavier Blasco Ferragud
Javier Sanchis Saez
Juan Manuel Herrero Durá

Acknowledgements

We would like to thank the departments and universities that hosted our research on multiobjective optimization over these years:

- *Instituto Universitario de Automática e Informática Industrial Universitat Politècnica de València*, Spain.
- *Programa de Pós-Graduação em Engenharia de Produção e Sistemas Pontificia Universidade Católica do Paraná*, Brazil.
- Spanish Ministry of Economy and Competitiveness with the projects: DPI2008-02133, TIN2011-28082, ENE2011-25900 and DPI2015-71443-R.
- National Council of Scientific and Technologic Development of Brazil (CNPq) with the postdoctoral fellowship BJT-304804/2014-2.

Also to our colleagues in this journey of the CPOH (http://cpoh.upv.es/): Sergio García-Nieto, Jesús Velasco, Miguel Martínez, José V. Salcedo, César Ramos and Raúl Simarro.

Contents

Acronyms

AOF	Aggregate Objective Function
DE	Differential Evolution
DM	Decision Maker
EA	Evolutionary Algorithm
EMO	Evolutionary Multiobjective Optimization
FEs	Function Evaluations
GFCL	Generate First, Choose Later
GPP	Global Physical Programming
IAE	Integral of the Absolute Value of Error
ITAE	Integral of the Time Weighted Absolute Value of Error
LD	Level Diagrams
MCDM	Multicriteria Decision Making
MIMO	Multiple Input, Multiple Output
MODE	Multiobjective Differential Evolution
MOEA	Multiobjective Evolutionary Algorithm
MOO	Multiobjective Optimization
MOOD	Multiobjective Optimization Design
MOP	Multiobjective Problem
PI	Proportional-Integral
PID	Proportional-Integral-Derivative
SISO	Single Input, Single Output
sp-MODE	Multiobjective Differential Evolution with Spherical Pruning

Part I
Fundamentals

This part is devoted to covering the *motivational and theorical* background required for this book on MOOD procedures for controller tuning. Firstly the motivation of this book will be provided, trying to answer the question *why multiobjective optimization techniques are valuable for controller tuning applications?* Afterwards, desirable features regarding the Multiobjective Problem (MOP) definition, the Evolutionary Multiobjective Optimization (EMO) process and the Multicriteria Decision Making (MCDM) stage, will be discussed. Finally, tools for the EMO process and the MCDM stage (used throughout this book) will be provided for practitioners.

Chapter 1
Motivation: *Multiobjective Thinking* in Controller Tuning

Abstract Throughout this chapter, we intend to provide a *multiobjective awareness* of the controller tuning problem. Beyond the fact that several objectives and requirements must be fulfilled by a given controller, we will show the advantages of considering this problem in its multiobjective nature. That is, optimizing simultaneously several objectives and following a multiobjective optimization design (MOOD) procedure. Since the MOOD procedure provides the opportunity to obtain a set of solutions to describe the objectives trade-off for a given multiobjective problem (MOP), it is worthwhile to use it for controller tuning applications. Due to the fact that several specifications such as time and frequency requirements need to be fulfilled by the control engineer, a procedure to appreciate the trade-off exchange for complex processes is useful.

1.1 Controller Tuning as a Multiobjective Optimization Problem: A Simple Example

Most engineering design statements, particularly controller tuning, can be formulated as an optimization problem. Firstly, the design problem has to be defined; that is to identify decision variables $\boldsymbol{\theta} = [\theta_1, ..., \theta_n]$ and design objectives $\boldsymbol{J} = [J_1(\boldsymbol{\theta}), \cdots, J_m(\boldsymbol{\theta})]$. For instance, consider a standard Proportional-Integral (PI) controller:

$$u(t) = K_c \left(e(t) + \frac{1}{T_i} \int_0^t e(t)dt \right) \tag{1.1}$$

For the PI tuning problem, the decision variables will be its parameters: the proportional gain K_c and the integral time T_i, that is $\boldsymbol{\theta} = [K_c, T_i]$.

The design objective may be a single index ($m = 1$). Assume for example that the *Integral of the Absolute Error* (IAE), the cumulative difference between desired and controlled output, is selected:

G. Reynoso Meza et al., *Controller Tuning with Evolutionary Multiobjective Optimization*, Intelligent Systems, Control and Automation: Science and Engineering 85, DOI 10.1007/978-3-319-41301-3_1

$$J_1(\boldsymbol{\theta}) = IAE(\boldsymbol{\theta}) = \int_{t=t_0}^{t_f} |r(t) - y(t)|\, dt = \int_{t=t_0}^{t_f} |e(t)|\, dt. \tag{1.2}$$

Then the tuning problem could be formulated as:

$$\begin{aligned} &\min_{\theta} J_1(\boldsymbol{\theta}) = \min_{\theta} IAE(\boldsymbol{\theta}) \\ st:& \\ &\underline{\theta_i} \leq \theta_i \leq \overline{\theta_i}, \quad i = [1, 2] \end{aligned} \tag{1.3}$$

where $\underline{\theta_i}$ and $\overline{\theta_i}$ are the lower and upper bounds of the decision variables.

Clearly, the solution obtained and its performance depends strongly on the design objective. For the following *first order plus time delay* model (delay and time constant in seconds),

$$\frac{Y(s)}{U(s)} = P(s) = \frac{3.2}{10s+1} e^{-3s} \tag{1.4}$$

the PI controller (see Fig. 1.1),

$$\frac{U(s)}{E(s)} = C(s) = K_c \left(1 + \frac{1}{T_i}\frac{1}{s}\right) \tag{1.5}$$

and the following constraints for decision variables $\underline{\boldsymbol{\theta}} = [0.1, 1]$ and $\overline{\boldsymbol{\theta}} = [10, 100]$, the optimal solution with the IAE as design objective is given by: $K_c = 0.640$ and $T_i = 10.68$ s. For this solution, the minimum IAE achieved is $J_1^{min} = IAE^{min} = 12.517$ Units·s. (see Fig. 1.2)

If a different objective is set, for instance $J_2(\boldsymbol{\theta}) = t_{98\,\%}(\boldsymbol{\theta})$, the time output y takes to get within 2 % of its final value, the new optimization problem will be:

$$\begin{aligned} &\min_{\theta} J_2(\boldsymbol{\theta}) = \min_{\theta} t_{98\,\%}(\boldsymbol{\theta}) \\ st:& \\ &\underline{\theta_i} \leq \theta_i \leq \overline{\theta_i}, \quad i = [1, 2] \end{aligned} \tag{1.6}$$

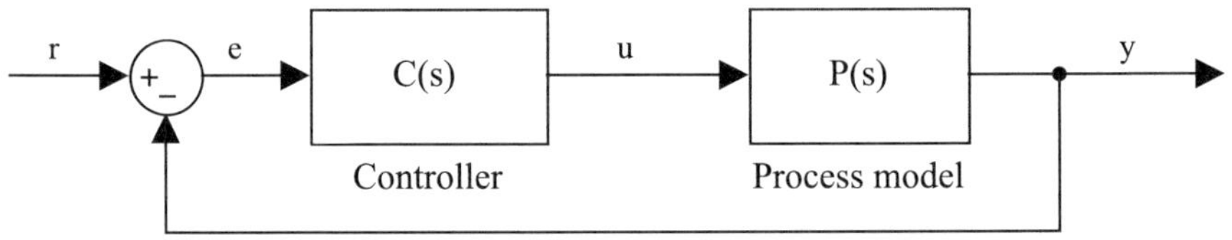

Fig. 1.1 Basic PI control structure

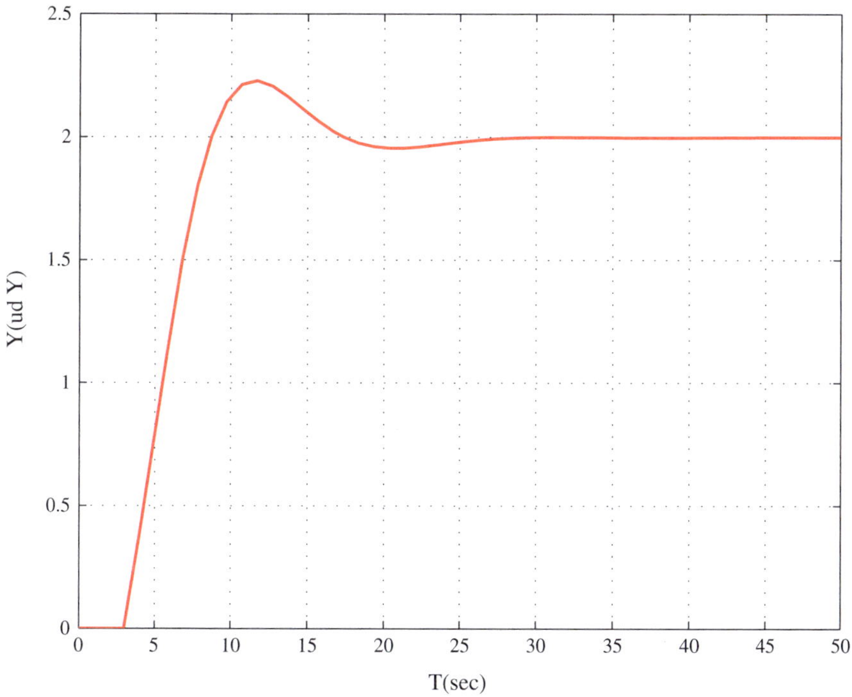

Fig. 1.2 Closed loop simulation with PI parameters obtained for IAE minimization. $r(t) = 2$

and the optimal solution now is: $K_c = 0.5444$ and $T_i = 11.08$ s, producing the minimum settling time $J_2^{min} = t_{98\,\%}^{min} = 10.6$ s. (Fig. 1.3 shows the simulation results and compares with the IAE solution obtained previously).

Table 1.1 and Fig. 1.4 compare the results obtained by each solution with respect to $J_1(\boldsymbol{\theta})$ and $J_2(\boldsymbol{\theta})$. Both solutions are *optimal* according to the objective for which they were calculated, but not when they are checked against the other. In general, this situation is common in a wide variety of engineering design problems; depending on the objective selected the solution may differ. This situation leads to the following questions:

- *Which solution is better?*
- *Which controller tuning should be implemented for the given process?*

Both answers lie in the practical aspects of the problem to solve. There is not a better solution than the other, but a solution with different trade off between (apparently) conflicting objectives. At the end, controller parameters K_c, T_i to be implemented will depend on the designer's preferences and the requirements that must be fulfilled for the given process. If one of the solutions fulfills the designer's expectations, then there is nothing more to be done and the tuning problem is solved by implementing the set of parameters from one of the above stated optimization problems.

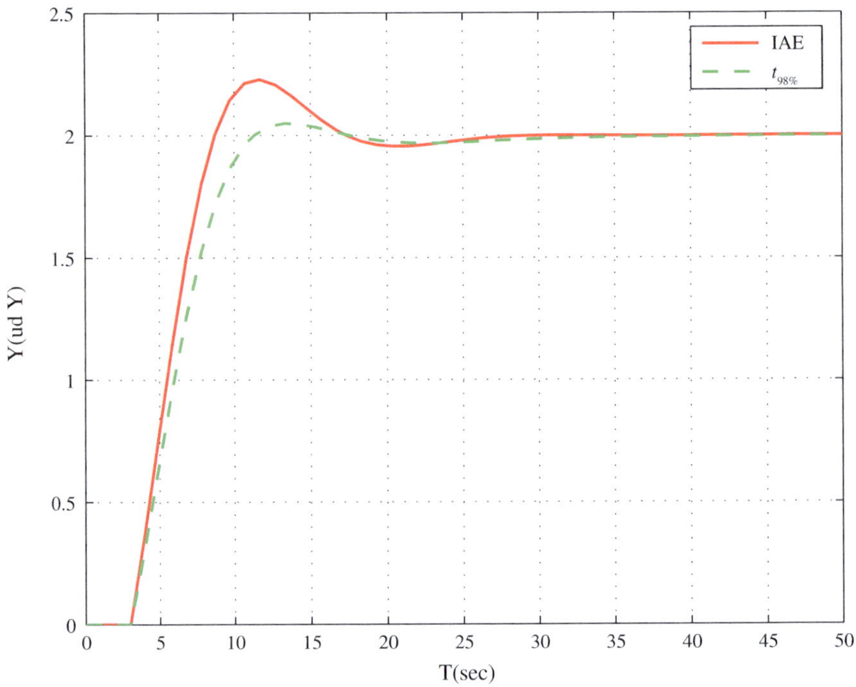

Fig. 1.3 Closed loop simulation with PI parameters obtained for $t_{98\,\%}$ minimization compared with the IAE minimization one. $r(t) = 2$

Table 1.1 Comparison of IAE and $t_{98\,\%}$ minimization results

Objective minimized	Decision variables $\boldsymbol{\theta}$	$J_1(\boldsymbol{\theta})$ *Units*·s	$J_2(\boldsymbol{\theta})$ s
IAE	$K_c = 0.6400$ $T_i = 10.68s$ s	12.517	22.3
$t_{98\,\%}$	$K_c = 0.5444$ $T_i = 11.08$ s	12.999	10.6

Nevertheless, the designer may be interested in the minimization of both objectives $J_1(\boldsymbol{\theta})$ and $J_2(\boldsymbol{\theta})$ simultaneously; that is, a solution with different exchange between those conflicting objectives. Figure 1.4 shows a wide area between both solutions where you could find new ones with different trade-offs. In that case, we are facing a multiobjective optimization problem (MOP), stated as:

$$\min_{\boldsymbol{\theta}} \boldsymbol{J} = [J_1(\boldsymbol{\theta}), \quad J_2(\boldsymbol{\theta})] \tag{1.7}$$

$$st:$$

$$\underline{\theta_i} \leq \theta_i \leq \overline{\theta_i}, \quad i = [1, 2].$$

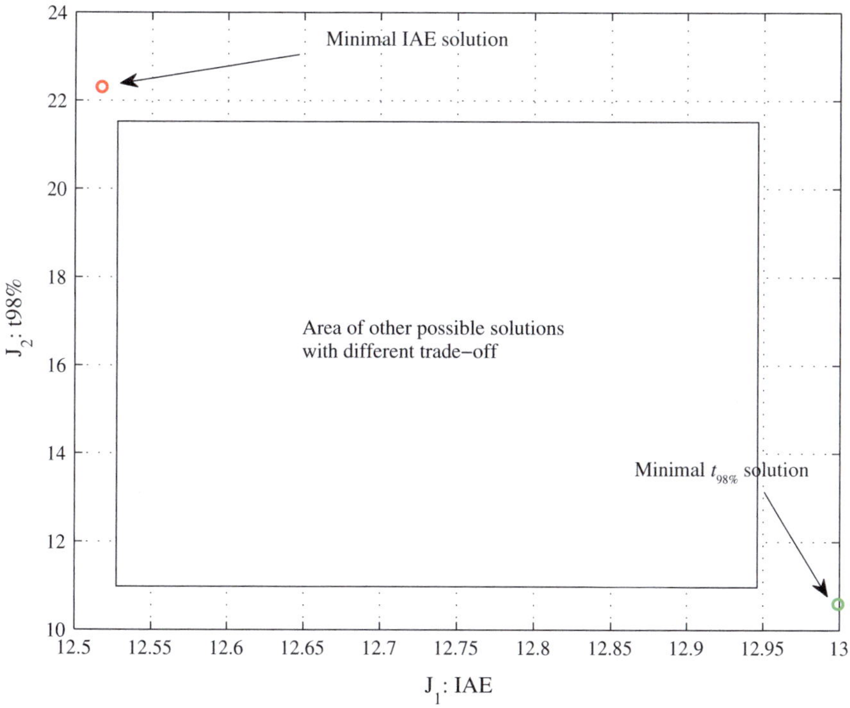

Fig. 1.4 Comparison of minimum IAE and minimum settling time solutions in the objective space

As more than one objective is set and the objectives are in conflict, there is no unique optimal solution but a set of optimal solutions (no one is better than any other). This optimal set is known as a **Pareto Set** and the objective values for this set comprises a **Pareto Front**.

Let's put the above commented thoughts within a general engineering design context, which will be solved through an optimization statement. Let m be the number of objectives in which the designer is interested. If $m = 1$, it is said to deal with a single-objective problem (Eqs. 1.3 or 1.6), while with $m \geq 2$ (Eq. 1.7) it is a MOP. In Fig. 1.5, a general (maybe brief) methodology to solve an engineering design problem by means of an optimization statement is presented.

According to [9], there are two main approaches to face a MOP: the *Aggregate Objective Function* (AOF) and the *Generate-First Choose-Later* (GFCL). In the AOF context, a mono-index optimization statement merging all the design objectives is defined. For instance, taking into account *time* and IAE as the *Integral of the Time weighted Absolute Error* (ITAE) is an easy way to apply AOF to our PI tuning example:

$$J_3(\boldsymbol{\theta}) = ITAE(\boldsymbol{\theta}) = \int_{t=t_0}^{t_f} t\,|r(t) - y(t)|\,dt. \tag{1.8}$$

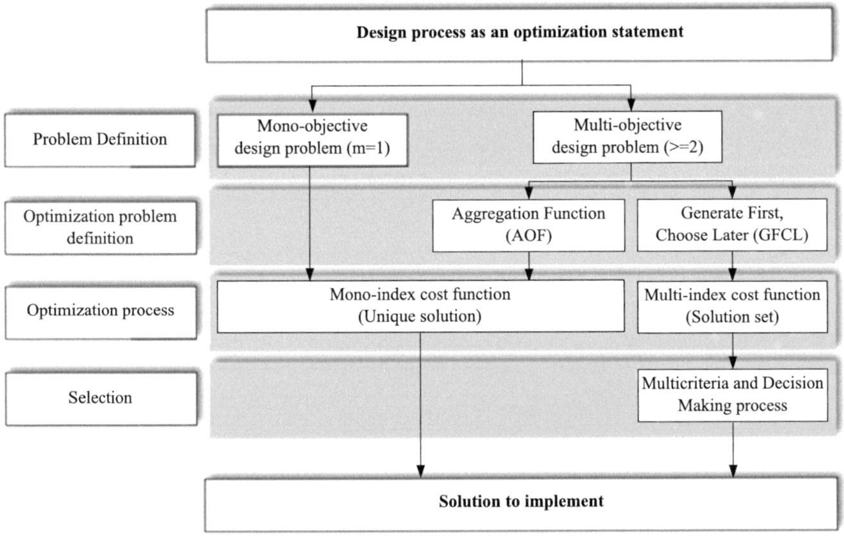

Fig. 1.5 Design methodology by means of optimization

Then a new optimization problem can be stated as:

$$\min_{\theta} J_3(\boldsymbol{\theta}) = \min_{\theta} ITAE(\boldsymbol{\theta}) \tag{1.9}$$
$$st:$$
$$\underline{\theta_i} \leq \theta_i \leq \overline{\theta_i}, \quad i = [1, 2]$$

and another new solution is obtained, which can be compared with previous solutions for IAE and $t_{98\,\%}$ minimization (Fig. 1.6 and Table 1.2). As expected, ITAE solution is the best for ITAE minimization but it would be an intermediate solution if preferred objectives were J_1 and J_2. As it is shown in Fig. 1.7, if ITAE solution is compared with IAE one, it is better in J_2 but worst in J_1 objective. Again, if ITAE solution is compared with $t_{98\,\%}$ one, it has better J_1 but worst J_2. For this situation, it is said that no one dominates the others (notice that none of the solutions are better in both objectives J_1 and J_2 simultaneously).

ITAE index is a traditional AOF which takes into account error and time, however it is difficult to know *a priori* which will be the trade-off between them. When the designer needs a different trade-off between objectives an intuitive AOF approach is by adding J_1 and J_2, using weights to express relative importance among them as in Eq. (1.10).

$$J_4(\boldsymbol{\theta}) = \alpha \cdot J_1(\boldsymbol{\theta}) + (1 - \alpha) \cdot J_2(\boldsymbol{\theta}). \tag{1.10}$$

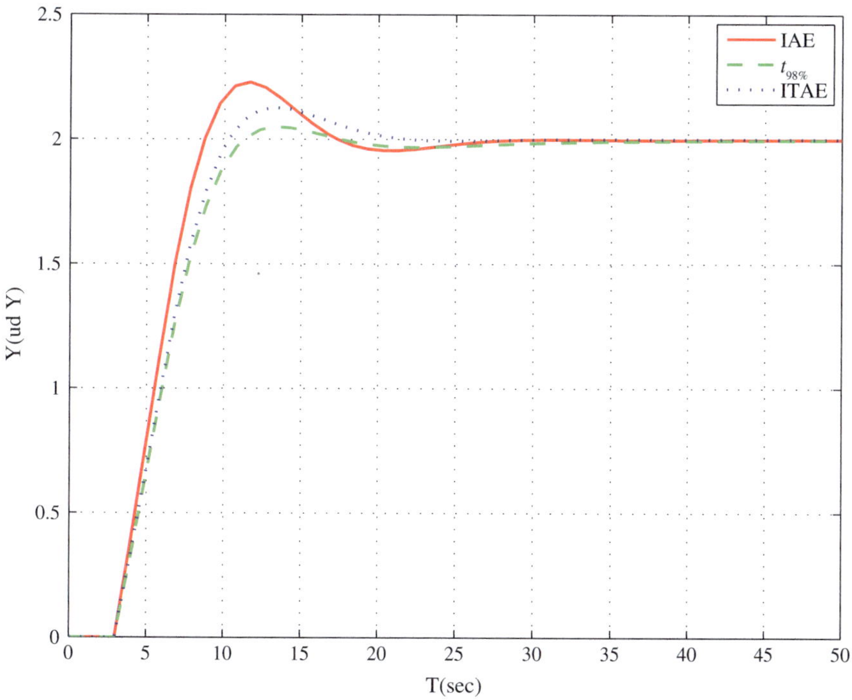

Fig. 1.6 Closed loop simulation with PI parameters obtained for ITAE minimization compared with the IAE and $t_{98\,\%}$ minimization ones

Table 1.2 Comparison of IAE, $t_{98\,\%}$ and ITAE minimization results

Objective minimized	Decision variables	J_1	J_2	J_3
		Units·s	s	*Units*·s^2
IAE (J_1)	$K_c = 0.6400$ $T_i = 10.68$ s	12.517	22.3	55.65
$t_{98\,\%}$ (J_2)	$K_c = 0.544$ $T_i = 11.08$ s	12.999	10.6	61.67
ITAE (J_3)	$K_c = 0.5512$ $T_i = 10.06$ s	12.778	17.8	51.26

With this formulation the designer has the possibility to assign, for instance, an 80 % of importance to J_1 and a 20 % to J_2 just setting $\alpha = 0.8$. But unfortunately achieving the desired balance between the two objectives also depends on other factors. The main problem probably is:

- *How to weight different objectives? or for this case ...*
- *How to equally compare IAE [Unit·s] and* $t_{98\,\%}$ *[s]?*

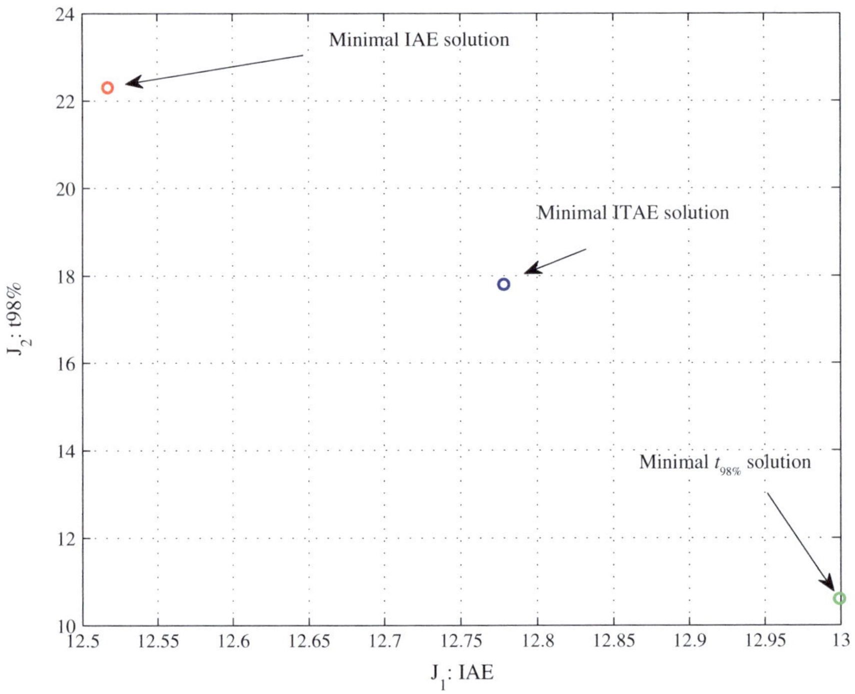

Fig. 1.7 Comparison of minimal IAE, ITAE and setling time solutions in the objective space

Therefore some normalization between objectives is needed to achieve a fair comparison among them. For example, it is popular in problems stated in terms of benefits and costs to put prices in terms of money, getting a common framework to compare design objectives. But in general for controller tuning is not always possible and the designer has to decide how to normalize and assign a physical or interpretable meaning. For our problem, IAE could be first dividing by the span time used in the integral; $T_{span} = t_f - t_0$, Eq. (1.11) to achieve some physical meaning (average error in units of the controlled variable).

$$J_1'(\boldsymbol{\theta}) = \frac{J_1(\boldsymbol{\theta})}{T_{span}}. \tag{1.11}$$

Besides, declaring which objective values are equivalent to allow comparison between their magnitudes/units will be required, for instance, according to a maximum value. With these considerations a new objective that aggregates the two original ones will be:

$$J_4(\boldsymbol{\theta}) = \alpha \frac{J_1'(\boldsymbol{\theta})}{J1'_{max}} + (1-\alpha) \cdot \frac{J_2(\boldsymbol{\theta})}{t_{max}}. \tag{1.12}$$

Table 1.3 Pareto optimal solutions obtained for different α values at J_4 minimization

Pareto solutions		Designer preference	Pareto front solutions	
K_c	T_i	α	$J_1(\boldsymbol{\theta})$	$J_2(\boldsymbol{\theta})$
0.5444	11.08	0	13.999	10.6
0.5444	11.08	0.05	12.999	10.6
0.5444	11.08	0.10	12.999	10.6
0.5444	11.08	0.15	12.999	10.6
0.5444	11.08	0.20	12.999	10.6
0.5444	11.08	0.25	12.999	10.6
0.5444	11.08	0.30	12.999	10.6
0.5444	11.08	0.35	12.999	10.6
0.5444	11.08	0.40	12.999	10.6
0.5444	11.08	0.45	12.999	10.6
0.5444	11.08	0.50	12.999	10.6
0.5444	11.08	0.55	12.999	10.6
0.5444	11.08	0.60	12.999	10.6
0.5444	11.08	0.65	12.999	10.6
0.5444	11.08	0.70	12.999	10.6
0.5444	11.08	0.75	12.999	10.6
0.5444	11.08	0.80	12.999	10.6
0.5444	11.08	0.85	12.999	10.6
0.5481	11.17	0.90	13.004	10.6
0.6288	10.56	0.95	12.520	16.1
0.6400	10.68	1.00	12.517	22.3

Notice that aggregation of several objectives, showing accurately the designer preferences, is not an easy task. In this case, the decision maker (DM or simply the designer) needs to describe the trade-offs at the beginning of the optimization process. Therefore, depending on the selected α, different results will be obtained (Table 1.3).

It can be observed in Table 1.3 that for several different values of α the same solution is obtained. Using a weight of $\alpha = 0.05$ (DM gives a 5 % of importance to J_1 and 95 % to J_2) is equivalent to minimizing with $\alpha = 0.9$ (90 % of importance to J_1 and 10 % to J_2). If we plot the results, as in Fig. 1.8, a wide area between groups of solutions and a high concentration of them in one extreme is observed. Notice that the ITAE solution is worst, regarding the two objectives, than the one obtained with $\alpha = 0.95$, because the ITAE uses a different way to agglutinate objectives; so this solution is optimal in the ITAE sense, but suboptimal in both objectives when comparing with the original design objectives. In fact, it seems there is not a lot of

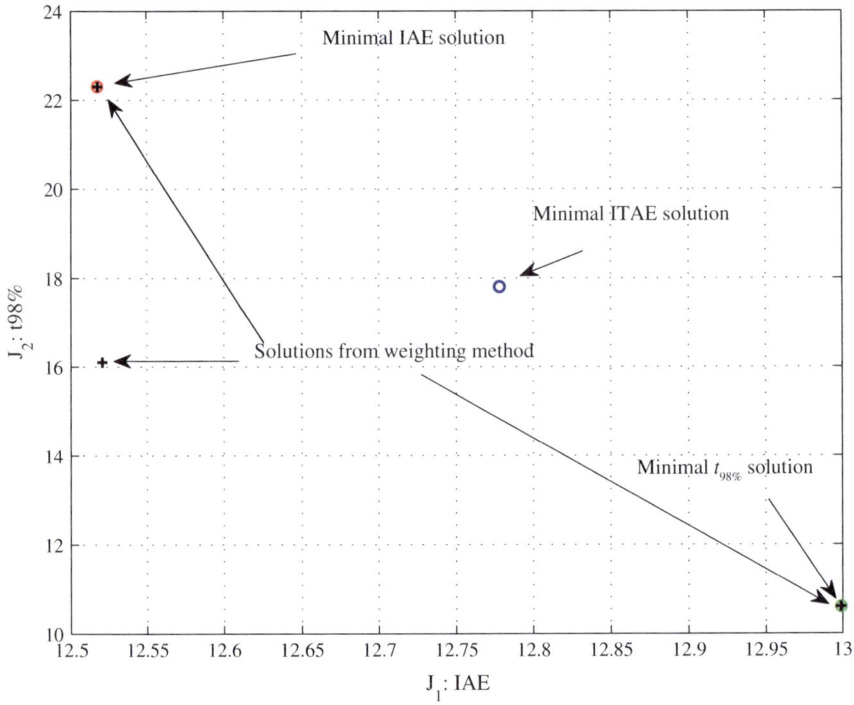

Fig. 1.8 Solutions for J_4 problem using different α values marked as +

options for the DM: minimum IAE, minimum $t_{98\,\%}$ or the middle solution presenting the same trade-off. This fact leads to the following questions:

- *Is the DM missing some information? or furthermore ...*
- *Is the weighting method an infallible way to specify trade-off between objectives?*

The answer is, it depends on the problem, but in general, it is a hard task to know *a priori* if the problem is going to be efficiently solved by the weighting method. The suitability of this method depends on the convexity and geometrical properties of the multiobjective problem. In our example, minimize IAE and settling time by the weighting method, shows efficiently a strong trade-off between design objectives; nevertheless, it does not seem a good alternative if the designer wants to sweep all possible trade-offs in order to analyze the solutions and select a preferable controller. **The first case represents the essence of the AOF approach and the second one, the essence of the GFCL method.**

When a better understanding of objectives trade-off is needed, multiobjective optimization may provide the required insight. For this purpose, a multiobjective optimization algorithm is needed to search for a good approximation of the Pareto front (without any subjective weighting selection). This optimization approach seeks for a set of Pareto optimal solutions to approximate the Pareto Front [8, 11]. This approximation provides a preliminary idea of an objective space; and according to [1], it could be helpful when it is necessary to explain and justify the MCDM procedure. As drawbacks, more time and embedment from the DM in the overall process is needed.

If a multiobjective optimization algorithm (further explained in Chap. 3) is used to solve our MOP (Eq. 1.7), the Pareto Front approximation of Table 1.4 and Fig. 1.9 is obtained. As it is shown, the front is non convex and presents discontinuities. **This kind of Pareto Front are difficult to approximate correctly.**

When simultaneous objective optimization is performed, the designer is provided with a set of solutions with different trade-offs. Notice that solving a MOP concludes with the selection of the final solution, which takes place in a Multicriteria Decision Making (MCDM) step. That is when the DM analyzes the trade-off among objectives, and selects the best solution according to his/her preferences. For that purpose, several techniques [5] and visualization approaches [7] are available; however this step may not be a trivial task since the analysis and visualization complexity increases with the number of objectives. Sometimes decision making could be more time consuming than the optimization process itself [3], and requires tools to help the DM. Moreover, notice there is a high degree of subjectivity in this extremely important step.

To illustrate the last idea, Fig. 1.10 shows two alternatives. The first one is the selection of the Pareto front solution nearest to the *ideal* point ($\boldsymbol{J}^{ideal}$). The *ideal* point is an utopian point built with the minimum values of both objectives. It seems to be a default choice to start but it is not always a preferred choice. Remark that each objective has its own units and that using pure geometrical distance could create a distortion or at least a misunderstanding of this measure. The second one could be the nearest point to a certain preferred area. If the DM has an idea about some preferred limits for J_1 and J_2, a preferred area to look for can be defined. Depending on the shape of the Pareto Front, some points could be inside and the DM could refine the final choice with an additional preference (for instance, proximity to ideal point). If none of the Pareto Front points are inside, the DM choice could be the nearest to this area of interest. (See Fig. 1.11 to compare the two aforementioned alternatives for final controller selection.)

Additionally to objectives definition and final decision procedure, a suitable multiobjective optimization algorithm is required. **As it has been shown in the example the type of problem has to be considered in the algorithm selection.** Lots of alternatives are available for different types of problems but a rough classification could be between convex optimization and global optimization.

Sometimes it is difficult and costly to produce a well distributed Pareto Front with techniques based on convex optimization. As example, let's apply the *goal attainment* methodology[1] to our PI tuning example. As most of the classical methods, converting the multiobjective problem into several single objective problems is the shortcut to obtain the Pareto Front. Therefore, Problem 1.7 is converted into multiple single objective problem (SOP) as:

[1]For example, using the *fgoalattain* function from the Matlab© Optimization Toolbox.

Table 1.4 Pareto Set, and Front approximations for IAE and $t_{98\,\%}$ minimization

Pareto set		Pareto front	
K_c	T_i	J_1	J_2
0.5444	11.08	12.999	10.6
0.5501	11.19	12.996	13.5
0.5487	11.16	12.992	13.6
0.5507	11.18	12.985	13.7
0.5507	11.17	12.978	13.8
0.5517	11.17	12.969	13.9
0.5517	11.16	12.961	14.0
0.5532	11.16	12.948	14.1
0.5553	11.18	12.934	14.2
0.5562	11.17	12.923	14.3
0.5570	11.16	12.911	14.4
0.5582	11.14	12.891	14.5
0.5608	11.15	12.871	14.6
0.5620	11.13	12.852	14.7
0.5644	11.12	12.831	14.8
0.5664	11.11	12.810	14.9
0.5695	11.10	12.783	15.0
0.5733	11.09	12.755	15.1
0.5764	11.07	12.729	15.2
0.5799	11.05	12.704	15.3
0.5841	11.02	12.674	15.4
0.5877	10.98	12.649	15.5
0.5940	10.95	12.616	15.6
0.5984	10.90	12.593	15.7
0.6038	10.84	12.569	15.8
0.6124	10.77	12.543	15.9
0.6199	10.68	12.527	16.0
0.6275	10.58	12.521	16.1
0.6291	10.55	12.520	16.2
0.6291	10.56	12.520	20.9
0.6285	10.57	12.520	21.1
0.6313	10.56	12.519	21.4
0.6340	10.58	12.518	21.7
0.6362	10.63	12.517	22.0
0.6400	10.68	12.517	22.3

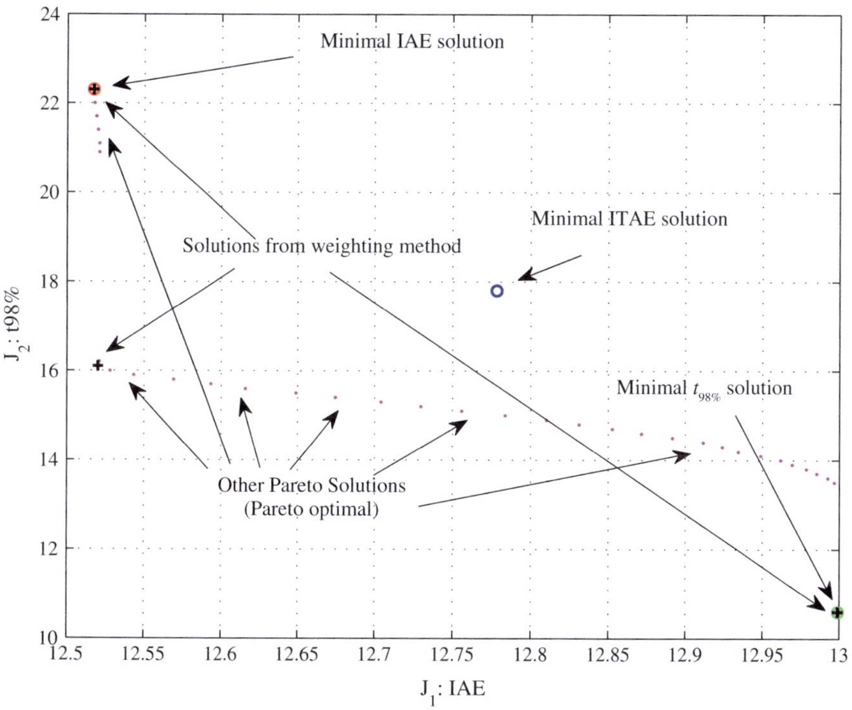

Fig. 1.9 Pareto Front approximation (•) for J_1 and J_2 minimization

$$\min_{\theta} \alpha \tag{1.13}$$

$$st:$$

$$\boldsymbol{J} = [J_1(\boldsymbol{\theta}), \quad J_2(\boldsymbol{\theta})] \tag{1.14}$$

$$\boldsymbol{J}^u + \alpha\boldsymbol{\omega} \geq \boldsymbol{J} \tag{1.15}$$

$$\underline{\theta_i} \leq \theta_i \leq \overline{\theta_i}, \quad i = [1, 2].$$

For each different $\boldsymbol{J}^u$ and/or different weighting vector $\boldsymbol{\omega}$ a different SOP is stated and solved by a convex optimization algorithm. Although the method is quite effective, several parameter adjustments ($\boldsymbol{J}^u$, $\boldsymbol{\omega}$) and initial points to feed the solver are needed and, if the problem is non-convex, a wrong starting point would produce a local optima or no solution at all.

For the selection of $\boldsymbol{J}^u$ and $\boldsymbol{\omega}$ (in charge of defining the different SOPs that will produce individual points of the Pareto Front) some knowledge of the problem is convenient. The extremes of the Pareto Front can be useful to bound the objective space area where the Pareto Front should be located. These extremes are obtained by minimizing J_1 and J_2 separately. Of course, if for this purpose a convex optimization

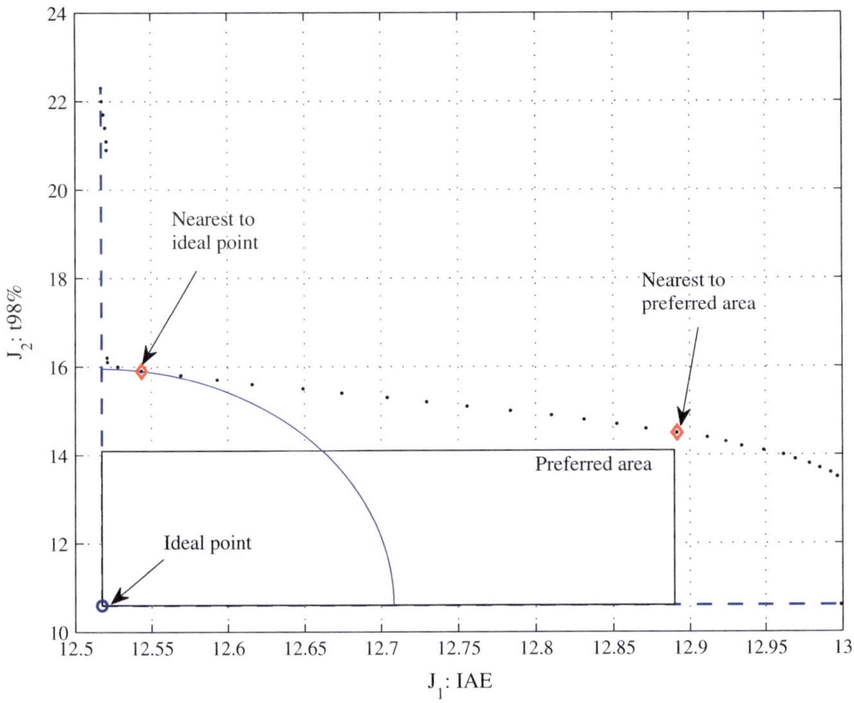

Fig. 1.10 Two alternatives for final selection step

algorithm[2] is used, it is necessary to supply an initial point and the designer has to guess or use *a priori* information to focus optimization algorithm into the proper area of the search space. Therefore minimum values of J_i (which produce the *ideal* point) can be an appropriate choice for the goal ($\boldsymbol{J}^u = \boldsymbol{J}^{ideal}$).

With a fixed goal, the weighting vector $\boldsymbol{\omega}$ changes the search direction and, ideally, it produces a different point of the front. For our PI tuning problem an intuitive way to select $\boldsymbol{\omega}$ could be changing the orientation from an angle $\beta = 0°$ to $\beta = 90°$ by increments according to the desired point distribution of the Pareto Front. Then, for a particular value of this angle β, the 2D weighting vector can be computed as:

$$\boldsymbol{\omega} = [\cos(\beta), \sin(\beta)]. \tag{1.16}$$

For each $\boldsymbol{\omega}$ stated, an initial point x^0 is needed to feed the algorithm which solves the SOP generated. After several trials, starting at random points and using the last result as the starting point of the next run, it is very difficult to achieve a solution similar to the one of Table 1.1. Other strategies exploiting the problem characteristics could be used: a starting point from a classical PID tuning methodology

[2]For instance, the *fmincon* function from the Matlab© Optimization Toolbox.

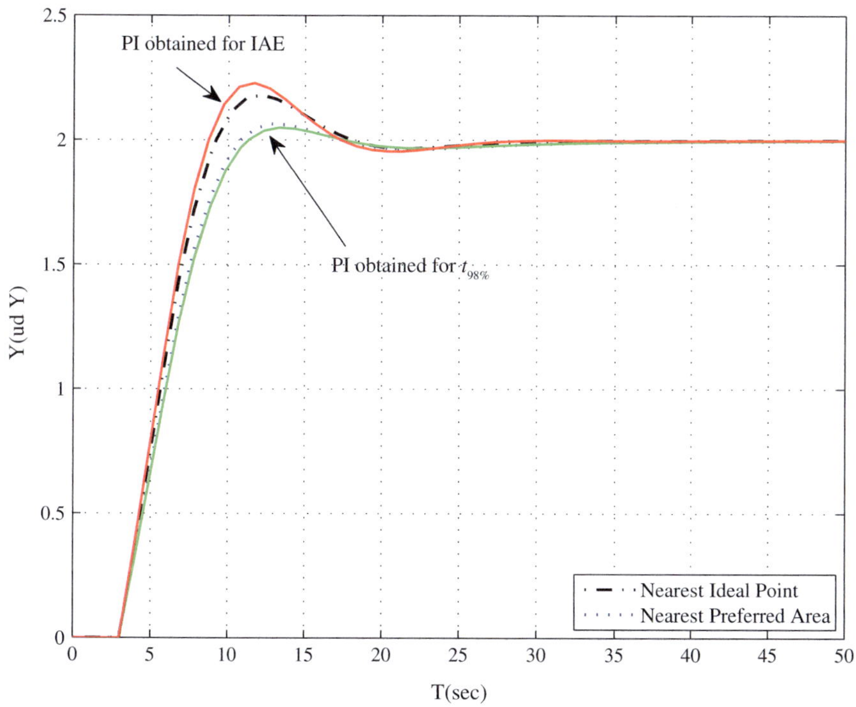

Fig. 1.11 Closed loop simulation with different PI selections: IAE, t98 %, nearest to ideal point and nearest to preferred area

(Ziegler-Nichols, S-IMC, etc.). Even so, some extra work exploiting the problem characteristics is necessary to help the solver if you want to succeed with the *goal attainment* method. Let's use the minima IAE and $t_{98\,\%}$ solutions obtained previously. It is reasonable (but not always true) that the Pareto solution should be inside an area between both minima solutions. Then a reasonable initial point should be in and for example, a linear distribution between minimum IAE and minimum $t_{98\,\%}$ solutions can be used. For instance, if 50 Pareto points are desired, initial guesses for the starting point x^0 and weighting vectors $\boldsymbol{\omega}$ can be calculated as:

$$
\begin{aligned}
\boldsymbol{\theta}_{IAE}^{min} &= [0.6400, 10.68] \\
\boldsymbol{\theta}_{t98\,\%}^{min} &= [0.5444, 11.18] \\
\boldsymbol{J}^{u} &= [J_1^{min} \quad J_2^{min}] \\
div &= 50,
\end{aligned}
$$

$$
\begin{aligned}
\beta_0 &= 0 \\
\beta_{step} &= 90^o/div
\end{aligned}
$$

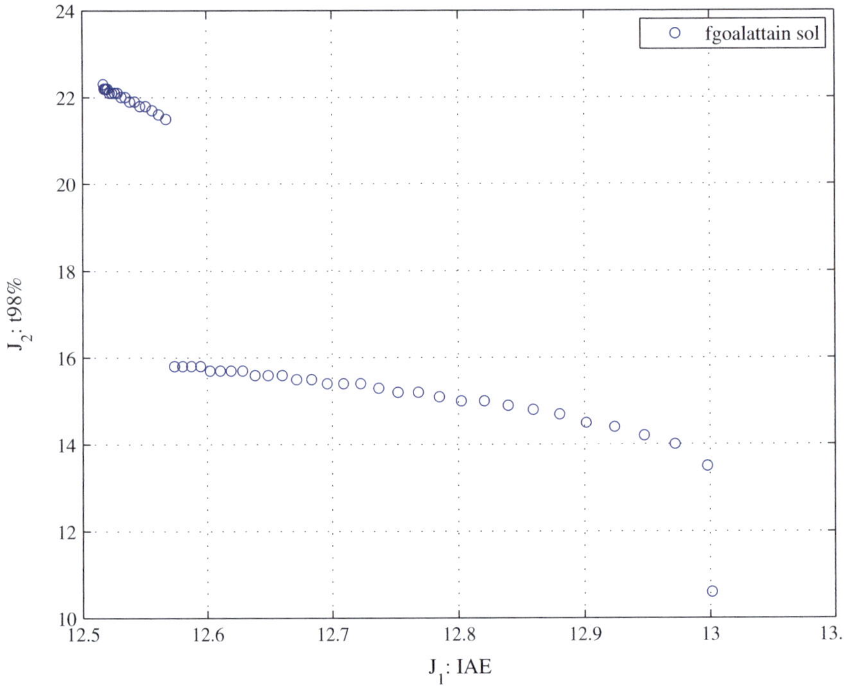

Fig. 1.12 Pareto Front approximation obtained by *goal attainment* procedure. In this case the *fgoalattain* from Matlab© has been used

$$\beta_k = \beta_{k-1} + \beta_{step} \quad k \in [1 \ldots div]$$
$$\boldsymbol{\omega} = [\cos(\beta_k), \sin(\beta_k)].$$

$$x^0 = \boldsymbol{\theta}^{min}_{t98\,\%}$$
$$x_{step} = \frac{\boldsymbol{\theta}^{min}_{IAE} - \boldsymbol{\theta}^{min}_{t98\,\%}}{div}$$
$$x^0_k = x^0_{k-1} + x_{step} \quad k \in [1 \ldots div].$$

With these selections it is possible to achieve the Pareto Front approximation shown in Fig. 1.12. Obviously, there is no guarantee that it was a good approximation but at least it is a candidate for it.

In summary, for our 2D problem, achieving this result had required several trial and error steps and additional work to help the optimizer. For a higher dimensional problem and time consuming objective functions, this process is more and more long and complex. It is fair to say that convex optimization has achieved a very high degree

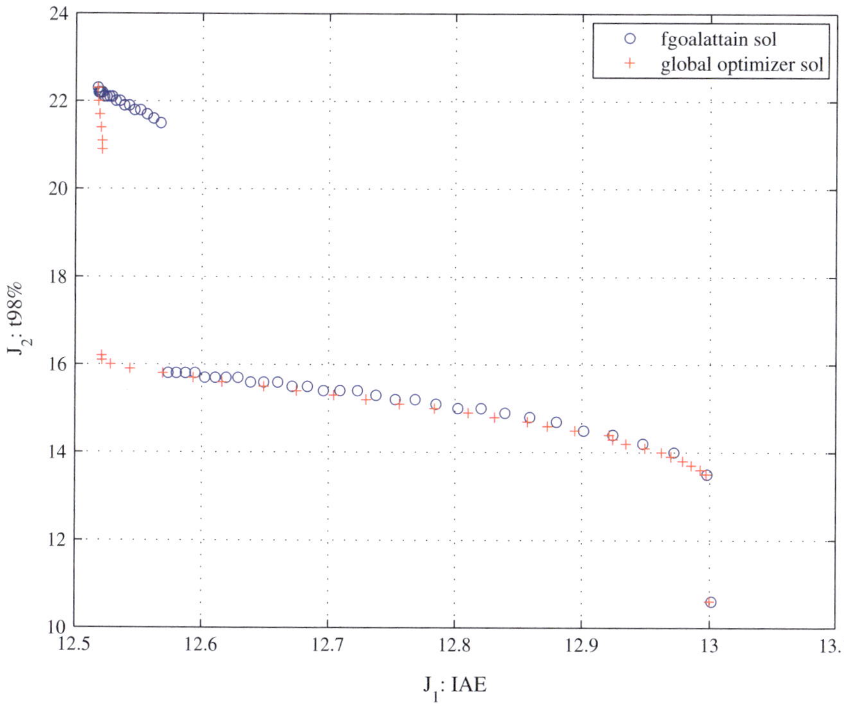

Fig. 1.13 Comparison of Pareto Front approximation obtained by *fgoalattain* function and a global optimizer

of development and can manage very effectively nonlinear problems with thousands of variables. It is highly recommended for a great variety of problems and it is easy to find commercial and open algorithms.

Traditionally, classic techniques [11] to calculate Pareto Front approximations have been used (such as varying weighting vectors, ϵ-constraint, and goal programming methods) as well as specialized algorithms (normal boundary intersection method [4], normal constraint method [10], and successive Pareto front optimization [13]).

But there is another set of problems where convex optimization is not enough and global optimization has to be used to increase the probability of achieving an accurate Pareto Front. For instance, if a global optimizer is used for our PI tuning problem the result shown in Fig. 1.13 produces a better front approximation than the one obtained by the *goal attainment* methodology. In this case, an evolutionary technique has been used (it will be described later). Since the computational cost is higher than the *goal attainment* procedure, a global optimizer increases the probability to obtain a better solution. An additional advantage of evolutionary techniques: it is not strictly necessary any previous tuning of the algorithm taking into account problem characteristics (for instance, initial points are not needed). So, when multiobjective problems are complex, nonlinear and highly constrained, a situation which makes it difficult to find a useful Pareto set approximation, another way to deal with MOPs is by means of Evolutionary Multiobjective Optimization (EMO), which is useful due to the flexibility of Multiobjective Evolutionary Algorithms (MOEAs) to deal with non-convex and highly constrained functions [2, 3]. Such algorithms have been successfully applied in several control engineering [6] and engineering design areas [14]. For this reason, MOEAs will be used in this book and hereafter the optimization process will be performed by means of EMO.

So far, in order to select a preferable set of parameters for our PI controller, following a multiobjective optimization approach, three fundamental steps were carried out:

1. Multiobjective problem (MOP) definition.
2. Multiobjective Optimization (MOO) process.
3. Multicriteria decision making (MCDM) step.

When the MOO process is merged with the MCDM step for a given MOP statement, it is possible to define a multiobjective optimization design (MOOD) procedure [12]. **This MOOD procedure can not substitute, in all instances, an AOF approach;** nevertheless, it could be helpful in complex design problems, where a close embedment of the designer is necessary. For example, where a trade-off analysis would be valuable for the DM before implementing a desired solution.

That is, in the case of controller tuning, the following questions should be answered:

- *Is it difficult to find a controller with a reasonable balance among design objectives?*
- *Is it worthwhile analyzing the trade-off among controllers (design alternatives)?*

If the answer to both is yes, then a MOOD procedure could fit into the controller tuning problem at hand.

1.2 Conclusions on This Chapter

In this chapter, some topics on MOP definitions, MOO process and MCDM step have been introduced. The aforementioned steps are important to guarantee the overall performance of a MOOD procedure. With a poor MOP definition, no matter how good the algorithms and MCDM methodology/tools are, the solutions obtained will not fulfill the DM's expectations. If the algorithm is inadequate for the problem at hand (regarding the desirable features 1–10), the DM will not obtain a useful Pareto set to analyze and therefore he/she will not be able to select a solution that meets his/her preferences. Finally, the wrong use of MCDM tools and methodologies could imply a lower degree of DM embedment in the trade-off analysis and the final selection. The last issue could easily discourage the DM from using MOOD procedure.

Regarding the MOP, some comments have been made about the capacity to reach a different level of interpretability on objective functions. In the MOOD approach there is no need to built a complicated aggregated function to merge the design objectives; therefore the objectives may be minded separately and optimized simultaneously. That is, the objective function statement could be done from the needs of the designer instead of the optimizer. This could facilitate the embedment of the designer in the overall procedure. In the case of MOO, it has been exposed how MOEAs could be useful to face different optimization instances as well as bring some desirable characteristics to the approximated Pareto front. It is important to remember that the final purpose of any MOEA is to provide the DM with a useful set of solutions (Pareto front approximation) to perform the MCDM procedure [1]. With regard to the MCDM step, notice that visualization of the Pareto front is a desirable tool to perform DM selections.

References

1. Bonissone P, Subbu R, Lizzi J (2009) Multicriteria decision making (MCDM): a framework for research and applications. IEEE Comput Intell Mag 4(3):48–61
2. Coello CAC, Lamont GB (2004) Applications of multi-objective evolutionary algorithms. In: Advances in natural computation, vol 1. World Scientific Publishing
3. Coello CAC, Veldhuizen DV, Lamont G (2002) Evolutionary algorithms for solving multi-objective problems. Kluwer Academic Press
4. Das I, Dennis J (1998) Normal-boundary intersection: a new method for generating the pareto surface in non-linear multicriteria optimization problems. SIAM J Optim 8:631–657
5. Figueira J, Greco S, Ehrgott M (2005) State of the art surveys. Springer international series. Multiple criteria decision analysis
6. Fleming P, Purshouse R (2002) Evolutionary algorithms in control systems engineering: a survey. Control Eng Practi 10:1223–1241
7. Lotov A, Miettinen K (2008) Visualizing the pareto frontier. In: Branke J, Deb K, Miettinen K, Slowinski R (eds) Multiobjective optimization vol 5252 of Lecture Notes in computer science. Springer, Berlin, pp 213–243
8. Marler R, Arora J (2004) Survey of multi-objective optimization methods for engineering. Struct Multidisciplinary Optim 26:369–395
9. Mattson CA, Messac A (2005) Pareto frontier based concept selection under uncertainty, with visualization. Optim Eng 6:85–115
10. Messac A, Ismail-Yahaya A, Mattson C (2003) The normalized normal constraint method for generating the pareto frontier. Struct Multidisciplinary Optim 25:86–98
11. Miettinen KM (1998) Nonlinear multiobjective optimization. Kluwer Academic Publishers
12. Reynoso-Meza G, Blasco X, Sanchis J (2012) Optimización evolutiva multi-objetivo y selección multi-criterio para la ingeniería de control. In: X Simposio CEA de Ingeniería de Control (March 2012), Comité Español de Automática, pp 80–87
13. Ruzika S, Wiecek M (2009) Successive approach to compute the bounded pareto front of practical multiobjective optimization problems. SIAM J Optim 20:915–934
14. Saridakis K, Dentsoras A (2008) Soft computing in engineering design - a review. Adv Eng Inf 22(2):202–221. Network methods in engineering

Chapter 2
Background on Multiobjective Optimization for Controller Tuning

Abstract In this chapter a background on multiobjective optimization and a review on multiobjective optimization design procedures within the context of control systems and the controller tuning problem are provided. Focus is given on multiobjective problems where an analysis of the Pareto front is required, in order to select the most preferable design alternative for the control problem at hand.

2.1 Definitions

A MOP, without loss of generality,[1] can be stated as follows:

$$\min_{\boldsymbol{\theta}} \boldsymbol{J}(\boldsymbol{\theta}) = [J_1(\boldsymbol{\theta}), \ldots, J_m(\boldsymbol{\theta})] \tag{2.1}$$

subject to:

$$\boldsymbol{g}(\boldsymbol{\theta}) \leq \boldsymbol{0} \tag{2.2}$$

$$\boldsymbol{h}(\boldsymbol{\theta}) = \boldsymbol{0} \tag{2.3}$$

where $\boldsymbol{\theta} \in D \subseteq \Re^n$ is defined as the decision vector in the searching space D, $\boldsymbol{J}(\boldsymbol{\theta}) \in \Lambda \subseteq \Re^m$ as the objective vector and $\boldsymbol{g}(\boldsymbol{\theta}), \boldsymbol{h}(\boldsymbol{\theta})$ as the inequality and equality constraint vectors, respectively. As remarked previously, there is no single solution to this problem because in general there is no best solution for all objectives. However, a set of solutions, the Pareto Set $\boldsymbol{\Theta}_P$, is defined, where each decision $\boldsymbol{\theta} \in \boldsymbol{\Theta}_P$ defines an objective vector $\boldsymbol{J}(\boldsymbol{\theta})$ in the Pareto Front. All solutions in the Pareto Front are said to be a set of Pareto-optimal and non-dominated solutions:

---

[1] A maximization problem can be converted to a minimization one. For each of the objectives to maximize, the transformation: $\max J_i(\boldsymbol{\theta}) = -min(-J_i(\boldsymbol{\theta}))$ should be applied.

G. Reynoso Meza et al., *Controller Tuning with Evolutionary Multiobjective Optimization*, Intelligent Systems, Control and Automation: Science and Engineering 85, DOI 10.1007/978-3-319-41301-3_2

Definition 2.1 (*Pareto Dominance*): A decision vector $\boldsymbol{\theta}^1$ dominates another vector $\boldsymbol{\theta}^2$, denoted as $\boldsymbol{\theta}^1 \preceq \boldsymbol{\theta}^2$, if $\boldsymbol{J}(\boldsymbol{\theta}^1)$ is not worse than $\boldsymbol{J}(\boldsymbol{\theta}^2)$ in all objectives and is better in at least one objective.

$$\forall i \in A := \{1, \ldots, m\}, \boldsymbol{J}_i(\boldsymbol{\theta}^1) \leq \boldsymbol{J}_i(\boldsymbol{\theta}^2) \wedge \exists i \in A : \boldsymbol{J}_i(\boldsymbol{\theta}^1) < \boldsymbol{J}_i(\boldsymbol{\theta}^2).$$

Definition 2.2 (*Strict Dominance* [91]): A decision vector $\boldsymbol{\theta}^1$ is strictly dominated by another vector $\boldsymbol{\theta}^2$ if $\boldsymbol{J}(\boldsymbol{\theta}^1)$ is worse than $\boldsymbol{J}(\boldsymbol{\theta}^2)$ in all objectives.

Definition 2.3 (*Weak Dominance* [91]): A decision vector $\boldsymbol{\theta}^1$ weakly dominates another vector $\boldsymbol{\theta}^2$ if $\boldsymbol{J}(\boldsymbol{\theta}^1)$ is not worse than $\boldsymbol{J}(\boldsymbol{\theta}^2)$ in all objectives.

Definition 2.4 (*Pareto optimal*): A solution vector $\boldsymbol{\theta}^*$ is Pareto optimal iff

$$\nexists \boldsymbol{\theta} \in D : \boldsymbol{\theta} \preceq \boldsymbol{\theta}^*.$$

Definition 2.5 (*Pareto Set*): In a MOP, the Pareto Set $\boldsymbol{\Theta}_P$ is the set including all the Pareto optimal solutions:

$$\boldsymbol{\Theta}_P := \{\boldsymbol{\theta} \in D | \nexists \boldsymbol{\theta}' \in D : \boldsymbol{\theta}' \preceq \boldsymbol{\theta}\}.$$

Definition 2.6 (*Pareto Front*): In a MOP, the Pareto Front $\boldsymbol{J}_P$ is the set including the objective vectors of all Pareto optimal solutions:

$$\boldsymbol{J}_P := \{\boldsymbol{J}(\boldsymbol{\theta}) : \boldsymbol{\theta} \in \boldsymbol{\Theta}_P\}.$$

For example, in Fig. 2.1, five different solutions $\boldsymbol{\theta}^1 \ldots \boldsymbol{\theta}^5$ and their corresponding objective vectors $\boldsymbol{J}(\boldsymbol{\theta}^1) \ldots \boldsymbol{J}(\boldsymbol{\theta}^5)$ are calculated to approximate the Pareto Set $\boldsymbol{\Theta}_P$

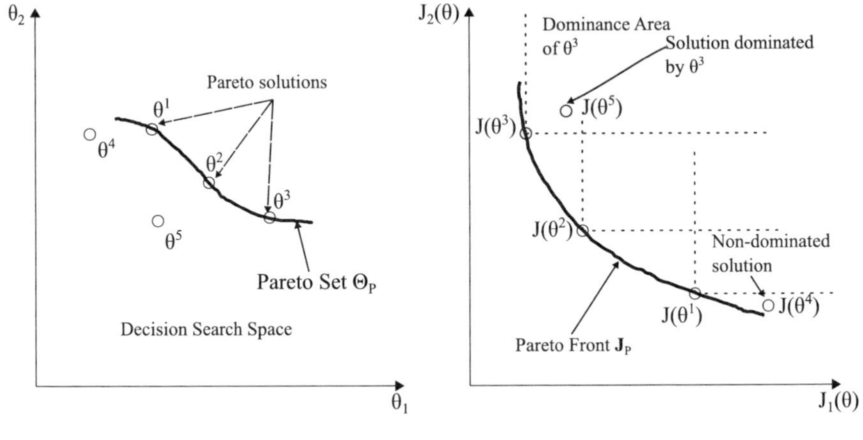

Fig. 2.1 Pareto optimality and dominance definitions. Pareto set and front in a bidimensional case ($m = 2$)

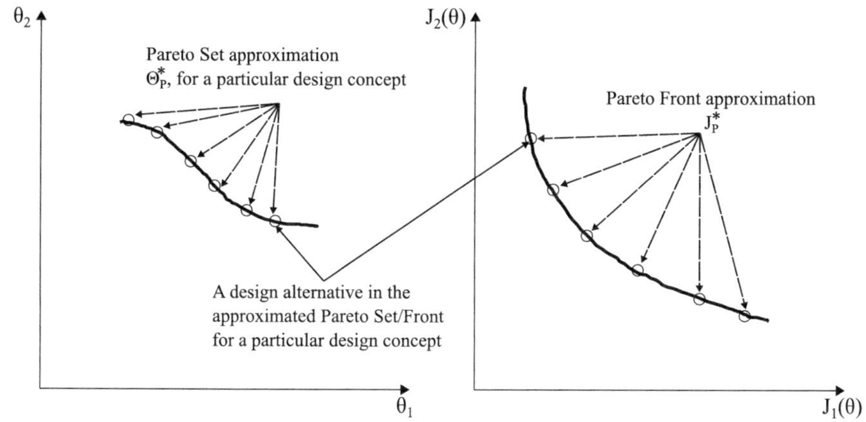

Fig. 2.2 Design concept and design alternative definitions

and Pareto Front $\boldsymbol{J}_P$ (bold lines). Solutions $\boldsymbol{\theta}^1 \ldots \boldsymbol{\theta}^4$ are non-dominated solutions, since there are no better solution vectors (in the calculated set) for all the objectives. Solution $\boldsymbol{\theta}^4$ is not Pareto optimal, since some solutions (not found in this case) dominate it. However, solutions $\boldsymbol{\theta}^1$, $\boldsymbol{\theta}^2$ and $\boldsymbol{\theta}^3$ are Pareto optimal, since they lie on the feasible Pareto front.

Obtaining $\boldsymbol{\Theta}_P$ is computationally infeasible, since most of the times the Pareto Front is unknown and likely it contains infinite solutions (notice that you shall only rely on approximations of the Pareto set $\boldsymbol{\Theta}_P^*$ and Front $\boldsymbol{J}_P^*$). In Fig. 2.1 the non-dominated solutions $\boldsymbol{\theta}^1 \ldots \boldsymbol{\theta}^4$ conform an approximated Pareto Set $\boldsymbol{\Theta}_P^*$ (although only $\boldsymbol{\theta}^1 \ldots \boldsymbol{\theta}^3$ are Pareto optimal) and their corresponding Pareto Front $\boldsymbol{J}_P^*$ approximation. Since $\boldsymbol{\Theta}_P$ contains all Pareto optimal solutions it will be desirable than $\boldsymbol{\Theta}_P^* \subset \boldsymbol{\Theta}_P$.

In [84], some refinement is incorporated into the Pareto Front notion to differentiate design concepts. A Pareto Front is defined given a design concept (or simply, a concept) that is, an idea about how to solve a given MOP. The design concept is built with a family of design choices (Pareto-optimal solutions) that are specific solutions in the design concept. In the leading example, the PI controller is the design concept, whereas a specific pair of proportional and integral gains is a design alternative. For example, in Fig. 2.2, the Pareto Set/Front (bold lines) for a particular design concept are approximated with a set of Pareto-optimal design alternatives (◯) (for example, a PI controller for a given MOP as a design concept). But sometimes, there are different concepts, all of which are viable for solving an MOP (for example, for a given control problem an LQR *or* a fuzzy controller can be used as alternative to PI concept). Therefore, the DM can calculate a Pareto Front approximation for each concept in order to make a comparison. Accordingly, in [84] the definition of Pareto Front and Pareto optimality were extended to a Pareto Front for a set of concepts (s-Pareto Front) where all solutions are s-Pareto optimal.

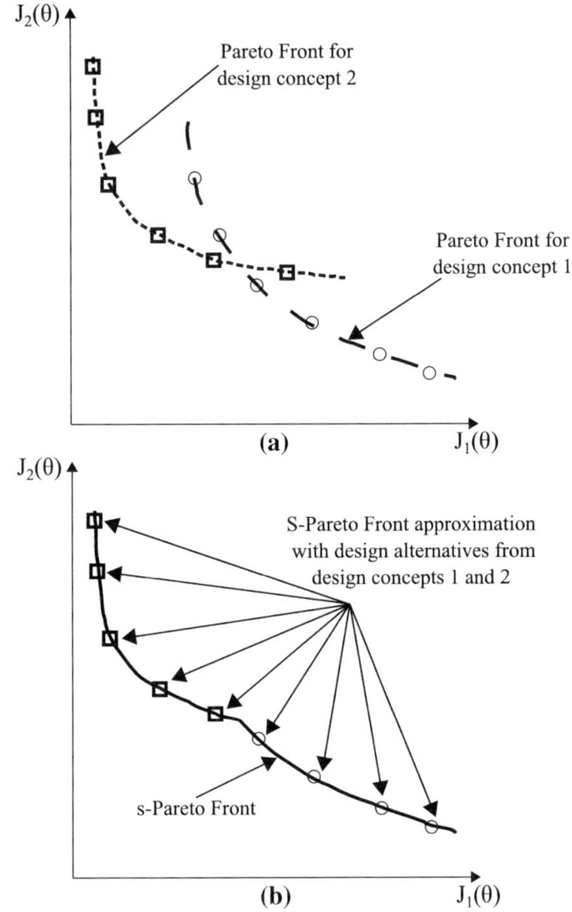

Fig. 2.3 s-pareto front definition

Definition 2.7 (*s-Pareto optimality*): Given an MOP and K design concepts, a solution vector $\boldsymbol{\theta}^1$ is s-Pareto optimal if there is no other solution $\boldsymbol{\theta}^2$ in the design concept k such that $J_i(\boldsymbol{\theta}^2) \leq J_i(\boldsymbol{\theta}^1)$ for all $i \in [1, 2, \ldots, m]$ and all concepts k, $k \in [1, \ldots, K]$; and $J_j(\boldsymbol{\theta}^2) < J_j(\boldsymbol{\theta}^1)$ for at least one j, $j \in [1, 2, \ldots, m]$ for any concept k.

Therefore, the s-Pareto Front is built joining the design alternatives of K design concepts. In Fig. 2.3 two different Pareto Front approximations for two different concepts (○ and □ respectively) are shown (Fig. 2.3a). In Fig. 2.3b, an s-Pareto Front with both design concepts is built.

As remarked in [84], a comparison between design concepts is useful for the designer, because he/she will be able to identify the concepts strengths, weaknesses, limitations and drawbacks. It is also important to visualize such comparisons, and to have a quantitative measure to evaluate these strengths and weaknesses.

In the next section, it will be discussed how to incorporate such notions into a design procedure for multiobjective problems.

2.2 Multiobjective Optimization Design (MOOD) Procedure

It is important to perform an entire procedure [9] minding equally the decision making and optimization steps [14]. Therefore, a general framework is required to successfully incorporate this approach into any engineering design process. In Fig. 2.4 a general framework for any MOOD procedure is shown. It consists of (at least) three main steps [18, 19]: the MOP definition (measurement); the multiobjective optimization process (search); and the MCDM stage (decision making).

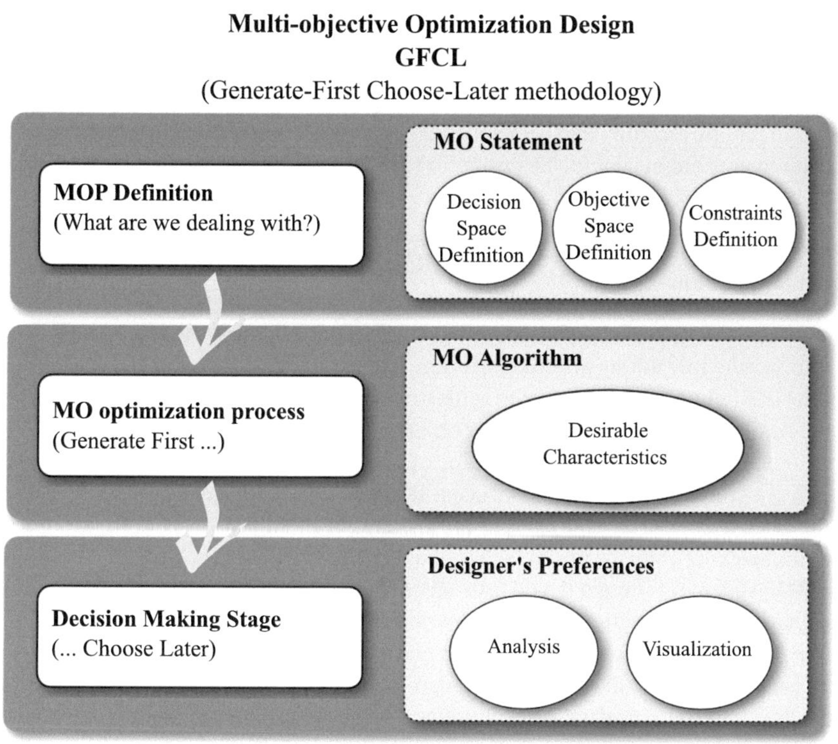

Fig. 2.4 A multiobjective optimization design (MOOD) procedure

2.2.1 Multiobjective Problem (MOP) Definition

In this stage, the design concept (how to tackle the problem at hand), the engineering requirements (which is important to optimize) and the constraints (which solutions are not practical/allowed) have to be defined. In [84] it is noted that the design concept implies the existence of a parametric model that defines the parameter values (the decision space) leading to a particular design alternative and performance (objective space). This is not a trivial task, since the problem formulation from the point of view of the designer is not that of the optimizer [45]. A lot of MOP definitions and their Pareto Front approximations have been proposed in several fields as described in [17]. Also, reviews on rule mining [123], supply chains [2, 79], energy systems [35, 38], flow shop scheduling [129], pattern recognition [21], hydrological modeling [34], water resources [107], machining [139], and portfolio management [88] can be consulted by interested readers.

The designer will search for a preferable solution at the end of the optimization process. As this book is dedicated to control system engineering, the discussed design concepts will be entirely related to this field. As a controller must satisfy a set of specifications and design objectives, a MOOD procedure could provide a deep insight into the controller's performance and capabilities. In counterpart, more time is required for optimization and decision making stages. Although several performance measurements are available, according to [3][2] the basic specifications will cover:

- Load disturbance rejection/attenuation.
- Measurement noise immunity/attenuation.
- Set point follow-up.
- Robustness to model uncertainties.

It is worthwhile noting how the selection of the optimization objectives for measuring the desired performance can be achieved. A convenient feature of using MOEAs is the flexibility to select interpretable objectives for the designer. That is, the objective selection can be close to the point of view of the designer. Sometimes, with classical optimization approaches, a cost function is built to satisfy a set of requirements such as convexity and/or continuity; that is, it is built from the point of view of the optimizer, in spite of a possible loss of interpretability for the designer. Therefore, the MOP statement is not a trivial task, since the problem formulation from the point of view of the designer is not that of the optimizer [45].

Given the MOP definition some characteristics for the MOEA could be required. That is, according to the expected design alternatives, the MOEA would need to include certain mechanisms or techniques to deal with the optimization statement. Some examples are related with robust, multi-modal, dynamic and/or computationally expensive optimization. Therefore, such instances could lead to certain desirable characteristics for the optimizer, which will be discussed in advance.

[2]Although specified in the context of PID control, they are applicable to all types of controllers.

2.2.2 Evolutionary Multiobjective Optimization (EMO)

Some of the classical strategies to approximate the Pareto Set/Front include: Normal constraint method [86, 116], Normal boundary intersection (NBI) method [24], Epsilon constraint techniques [91] and Physical programming [87]. In [55], a Matlab© toolbox kit for automatic control[3] is presented that includes some of the aforementioned utilities for multiobjective optimization. For the interested reader, in [81, 91] reviews of general optimization statements for MOP in engineering are given. However, as noticed earlier, this book focuses on the MOOD procedure by means of EMO so MOEAs will be discussed.

MOEAs have been used to approximate a Pareto set [144], due to their flexibility when evolving an entire population towards the Pareto front. A comprehensive review of the early stages of MOEAs is contained in [20]. There are several popular evolutionary and nature-inspired techniques used by MOEAs. The former, mainly based on the laws of natural selection where the fittest members (solutions) in a population (set of potential solutions) are more likely to survive as the population evolves. The latter is based on the natural behavior of organisms. Anyway in both cases a population is evolved towards the (unknown) Pareto Front. We will refer to them simply as evolutionary techniques.

The most popular techniques include Genetic Algorithms (GA) [69, 122], Particle Swarm Optimization (PSO) [15, 65], and Differential Evolution (DE) [27, 28, 90, 128]. Nevertheless, evolutionary techniques as Artificial Bee Colony (ABC) [64], Ant Colony Optimization (ACO) [33, 93] of Firefly algorithms [42] are becoming popular. No evolutionary technique is better than the others, since each has its drawbacks and advantages. These evolutionary/nature-inspired techniques require mechanisms to deal with EMO since they were originally used for single objective optimization. While the dominance criterion (Definition 2.1) could be used to evolve the population towards an approximated Pareto Front, it could be insufficient to achieve a minimum degree of satisfaction in other desirable characteristics for a MOEA (diversity, for instance). In Algorithm 2.1 a general structure for a MOEA is given. Its structure is very similar to most evolutionary techniques [43]: it builds and evaluates an initial population $P|_0$ (lines 1–2) and archives an initial Pareto Set approximation (line 3). Then, optimization (evolutionary) process begins with the lines 5–10. Inside this optimization process, the evolutionary operators (depending on the evolutionary technique) will build and evaluate a new population (line 7–8), and the solutions with better cost function will be selected for the next generation (line 10). The main difference is regarding line 9, where the Pareto Set approximation is updated; according to the requirements of the designer, such process will incorporate (or not) some desirable features.

Desirable characteristics for a MOEA could be related to the set of (useful) solutions required by the DM or the optimization design statement at hand (Fig. 2.5). Regarding a Pareto Set, some desirable characteristics include (in no particular order) convergence, diversity and pertinency. Regarding the optimization statement,

[3]Freely available at http://www.acadotoolkit.org/.

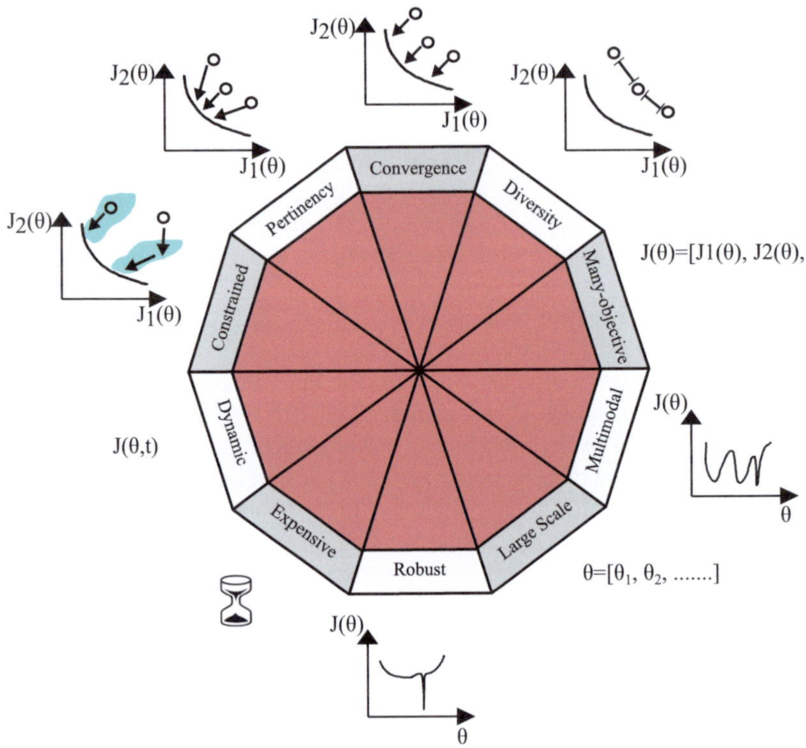

Fig. 2.5 Desired properties for MOEAs

Build initial population $P|_0$ with N_p individuals;
Evaluate $P|_0$;
Build initial Pareto set approximation $\boldsymbol{\Theta}^*_P|_0$ with $P|_0$;
Set generation counter $G = 0$;
while *convergence criteria not reached* **do**
  $G = G + 1$;
  Build population $P^*|_G$ using $P|_{G-1}$ with an evolutionary or bio-inspired technique;
  Evaluate new population $P^*|_G$;
  Build Pareto set approximation $\boldsymbol{\Theta}^*_P|_G$ with $\boldsymbol{\Theta}^*_P|_{G-1} \bigcup P^*|_G$;
  Update population $P|_G$ with $P^*|_G \bigcup P|_{G-1}$;
end
RETURN Pareto set approximation $\boldsymbol{\Theta}^*_P|_G$;

Algorithm 2.1: Basic MOEA.

some features could be related to deal with constrained, many-objectives, dynamic, multi-modal, robust, computationally expensive or large scale optimization instances. These desired characteristics are also a guide to appreciate current trends and on going research on EMO and MOEAs development [30, 144].

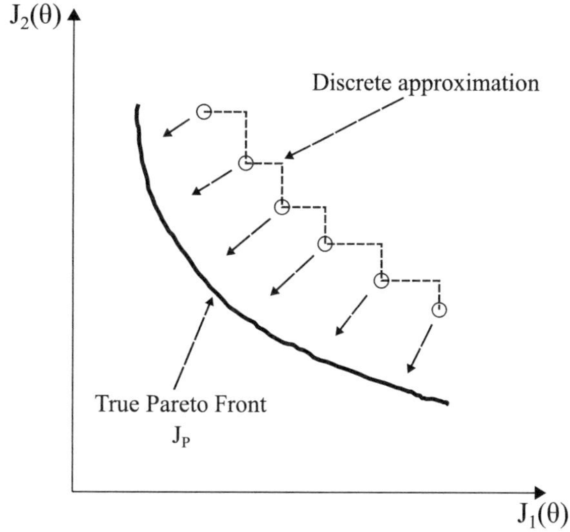

Fig. 2.6 Convergence towards the pareto front

Feature 1 *Convergence*

Convergence is the algorithm's capacity to reach the real (usually unknown) Pareto front (Fig. 2.6). Convergence properties usually depend on the evolutionary parameters of the MOEA used. Because of this, several adaptation mechanisms are available as well as several *ready to use* MOEAs with a default set of parameters. For example, the CEC (Congress on Evolutionary Computation) benchmarks on optimization [58, 142] provide a good set of these algorithms, comprising evolutionary techniques as GA, PSO and DE. Another idea to improve the convergence properties of a MOEA is by means of using local search routines through the evolutionary process. Such algorithms are know as memetic algorithms [95, 98].

Evaluating the convergence of a MOEA over another is not a trivial task, since you are comparing Pareto front approximations. For two objectives it could not be an issue, but in several dimensions is more difficult. Several metrics have been developed to evaluate the convergence properties (and other characteristics) for MOEAs [67, 148].

Convergence is a property common to all optimization algorithms; from the user's point of view it is an expected characteristic. Nevertheless, in the case of MOEAs it could be insufficient, and another desired (expected) feature, as diversity, is required.

Feature 2 *Diversity Mechanism*

Diversity is the algorithm's capacity to obtain a set of well-distributed solutions on the objective space; thus providing a useful description of objectives and decision variables trade-off (Fig. 2.7). Popular ideas include pruning mechanisms, spreading measures or performance indicators of the approximated front.

Regarding pruning mechanisms, probably the first technique was the ϵ-dominance method [70], which defines a threshold where a solution dominates other solutions

Fig. 2.7 Diversity notion in the pareto front

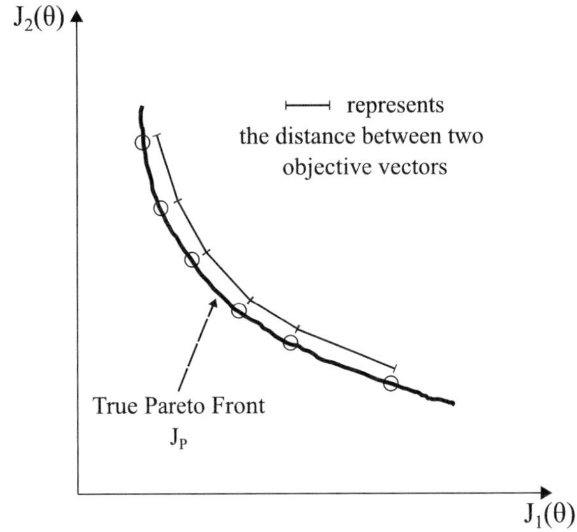

in their surroundings. That is, a solution dominates the solutions that are less fit for all the objectives, as well as the solutions inside a distance than a given parameter ϵ. Such dominance relaxation has been shown to generate Pareto Fronts with some desirable pertinency characteristics [82]. Algorithms based on such concept include ev-MOGA[4] [52], paϵ-MyDE [51], and paϵ-ODEMO [48]. Similar ideas have been developed using spherical coordinates (or similar statements) [5, 10, 113] in the objective or decision space.

In regard to spreading measures, the crowding distance [31] is used to instigate an algorithm to migrate its population to less crowded areas. This approach is used in algorithms such as NSGA-II[5] [31], which is a very popular MOEA. Other algorithms such as MOEA/D[6] [141] decompose the problem in several scalar optimization subproblems, which are solved simultaneously (as in NBI algorithm) and thereby assure diversity as a consequence of space segmentation when defining the scalar subproblems.

In the case of performance indicators, instead of comparing members of the population, at each generation solutions who best build a Pareto Front are selected based on some performance indicator. An example is the IBEA algorithm [147] which is an indicator based evolutionary algorithm. Most used performance indicators are the hypervolume and the epsilon-indicator [148].

However a good diversity across the Pareto Front must not be confused with solution pertinency (meaning, interesting and valuable solutions from the DM point of view). Several techniques trying to accomplish a good diversity on the Pareto Front

[4]Available for Matlab© at: http://www.mathworks.com/matlabcentral/fileexchange/31080.

[5]Source code available at: http://www.iitk.ac.in/kangal/codes.shtml; also, a variant of this algorithm is available in the global optimization toolbox of Matlab©.

[6]Matlab© code available at: http://cswww.essex.ac.uk/staff/zhang/IntrotoResearch/MOEAd.htm.

Fig. 2.8 Pertinency notion

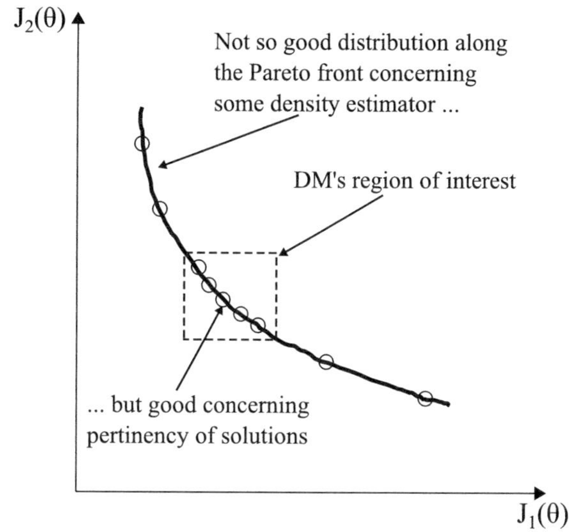

seem to be based on (or compared with) uniform distributions. Nevertheless, a large set of solutions may not be of interest to the DM, owing to a strong degradation in one (or several) objectives [22]. Therefore, some mechanisms to incorporate designer preferences could be desirable to improve solutions pertinency.

Feature 3 *Pertinency*

Incorporating DM preferences into a MOEA has been suggested in order to improve the pertinency of solutions. That is, improving the capacity to obtain a set of interesting solutions from the DM point of view (Fig. 2.8). Several ways to include designer's preferences in the MOOD procedure comprise *a priori*, *progressive*, or *a posteriori* methods [96].

- A priori: the designer has some knowledge about his/her preferences in the objective space. In such cases, you could be interested in an algorithm that could incorporate such preferences in the optimization procedure.
- Progressive: the optimization algorithm embeds the designer into the optimization process adjusting his/her preferences *on the fly*. This could be a desirable characteristic for an algorithm when the designer has some knowledge of the objectives trade-off in complex problems.
- A posteriori: the designer analyzes the Pareto Front calculated by the algorithm and, according to the set of solutions, he/she defines his/her preferences in order to select the preferable solution.

Some popular techniques include ranking procedures, goal attainment, fuzzy relations, among others [14]. Improving pertinency in multiobjective algorithms could have a direct and positive impact in the MCDM stage, since the DM could be provided with a more compact set of potential and interesting solutions. It has been suggested that the size of the Pareto Front approximation must be kept to a manageable size for the DM. According to [87] it is usually impossible to retain information from more than 10 or 20 design alternatives.

A natural choice to improve solutions' pertinency is the inclusion of optimization constraints (besides bound constraints on decision variables). This topic will be exposed below.

Feature 4 *Constrained optimization*

Another desirable characteristic in MOEAs is constraint handling. Since most of the design optimization problems need to consider constraints, such mechanisms are always an interesting topic of research. Various techniques have been developed for evolutionary optimization [16, 44]. In [89], those techniques are classified as:

- Feasibility rules. An easy and basic manner to implement the approach is discussed in [29]. It consists in:
  - When comparing two feasible solutions, the one with the best objective function is selected.
  - When comparing a feasible and an infeasible solution, the feasible one is selected.
  - When comparing two infeasible solutions, the one with the lowest sum of constraint violation is selected.
- Stochastic ranking. This approach briefly consists in comparing two infeasible solutions by their fitness or by their constraint violations.
- ϵ-constrained method. This method uses a lexicographic ordering mechanism where the minimization of the constraint violation precedes the minimization of the objective function. This mechanism, with an adaptive parameter scheme,[7] won the CEC2010 competition in a special session on constrained real-parameter optimization [77].
- Novel penalty functions and novel special operators.
- Multiobjective concepts. In the case of MOO, it can be a straightforward approach where the constraint is treated as an additional objective to optimize towards a desired value (goal vector).
- Ensemble of constraint-handling techniques. This approach involves taking advantage of all the mechanisms for constraint handling and using them on a single optimization run (for example [78]).

Regarding controller tuning, constrained optimization instances may appear in complex processes where, for example, several constraints on settling time, overshoot and robustness must be fulfilled.

[7]Code available at http://www.ints.info.hiroshima-cu.ac.jp/~takahama/eng/index.html for single objective optimization.

Feature 5 *Many-Objectives optimization*

Algorithms with good diversity preservation mechanisms could have problems if solutions are dominance resistant in an m-dimensional objective space and so waste time and resources in non-optimal areas [104]. This is because of the self diverse nature and the large number of objectives (usually, $m \geq 5$). Furthermore, recent research has indicated that a random search approach can be competitive for generating a Pareto Front approximation for a many-objectives optimization [22]. Several approaches to deal with many-objectives optimization include [61]:

- Modification of Pareto dominance to improve the selection pressure towards the Pareto Front.
- Introduction of different ranks to define a metric based on the number of objectives for which a solution is better than the other.
- Use of indicator functions as performance indicators of the quality of the Pareto Front approximation.
- Use of scalarizing functions (weighting vectors, for example).
- Use of preference information (see above), that is, information on the region of interest for the DM.
- Reduction in the number of objectives.

Examples to deal with this last issue can be seen in [75] where an objective reduction is performed using principal component analysis (PCA), or in [120] where a heuristic approach is used for dimensional reduction. Besides, algorithms which incorporate preference information in the optimization approach could be used in many-objective instances [61].

In the specific case of controller tuning, a many-objective optimization instance would appear according with the complexity of a given control loop or process, and the number of requirements to fulfill.

Feature 6 *Dynamic optimization*

Sometimes the static approach is not enough to find a preferable solution and therefore, a dynamic optimization statement needs to be solved where the cost function is varying with time. The challenge, besides tracking the optimal solution, is to select the desired solution at each sampling time. In [23, 36] there are extensive reviews on this topic.

As it can be noticed, this kind of capabilities would be useful for problems related with Model Predictive Control (MPC) where a new control value is obtained at each sampling time taking into account new information of the process outputs.

Feature 7 *Multi-modal Optimization*

Multi-modal instances for controller tuning *per se* seem to be not usual; nevertheless they may appear in multi disciplinary optimization [83] statements, where besides the tuning parameters, other design variables (as mechanical or geometrical) are involved.

In multi-modal optimization, different decision variable vectors could give the same objective vector. In some instances, it could be desirable to retain such solutions and perform, in the MCDM step, an analysis according with the decision space region where those solutions belongs. This could be important in instances where, for example, the decision variables have a physical meaning and it is convenient to analyze the impact of using one over another. In a EMO framework, this information could be added as additional objectives as noticed by [32]. For more details on multi-modal optimization, the interested reader could refer to [26].

Feature 8 *Robust Optimization*

In a general frame and according to [7], robust optimization could refer not only to the models used to measure the performance, but also with the sensitivity analysis on the calculated solutions. That is, how much could be degraded the objective vector under the presence of uncertainties. This sensibility analysis could be done by means of deterministic measures and/or with direct search (as Montecarlo methods). This kind of analysis could bring a different level of interpretability of the performance due to uncertainties in the model used in the optimization. This problem statement is related with reliability optimization, where a given performance must be assured for a certain solution along different scenarios.

An example is provided in [124] where an evaluation of the *American Control Conference Benchmark* [136] based on Montecarlo methods is done.

Feature 9 *Computationally Expensive optimization*

Computationally expensive optimization is related with line 8 of Algorithm 2.1. Sometimes cost function evaluation requires a huge amount of computational resources. Therefore stochastic approaches are a disadvantage, given the complexity to evaluate the fitness (performance) of an individual (design alternative). Recent solutions are mainly oriented to generate a surface *on-the-fly* of the objective space, with lower computational effort. One popular technique is the use of Neural Networks, trained through an evolutionary process, but any kind of model or surface approximation could be used. A review on the topic can be consulted in [117]. In the field of control systems engineering, such type of instances would appear when expensive calculations in complex simulations are needed to compute the objective vector.

In other instances, such computational effort could be relative; that is, there are limited computational resources to evaluate a cost function. To deal with this issue *compact evolutionary algorithms* has been proposed, but such instance has not reach yet the EMO approach. Some examples are exposed in [50] and [92]. Instances where this capabilities could be desirable include embedded solvers for optimization.

Feature 10 *Large scale optimization*

It refers to the capabilities of a given MOEA to deal with MOP with any number of decision variables with reasonable computational resources. Sometimes a MOEA can perform well for a relatively small number of decision variables, but it could be an impractical solution (according to the computational resources available) to solve a problem with a bigger number of decision variables. Whilst in expensive optimization instances (Feature 9) the complexity is due to the performance measurement (line 8 in Algorithm 2.1), in large scale may be related to the algorithm's mechanism to approximate a new set of design alternatives (lines 7 and 9). In the former the complexity is added by the problem, in the latter by the algorithm. A review on this topic can be consulted in [74].

The aforementioned features could be desirable characteristics for a given MOEA. Afer all, it would depend on the designer's preferences and the MOP statement at hand. Afterwards, a MCDM step must be carried, in order to select the most preferable solution. This step is commented below.

2.2.3 MultiCriteria Decision Making (MCDM)

Once the DM has been provided with a Pareto Front $\boldsymbol{J}_P^*$, she/he will need to analyze the trade-off between objectives and select the best solution according to her/his preferences. A comprehensive compendium on MCDM techniques (and software) for multi-dimensional data and decision analysis can be consulted in [41]. Assuming that all preferences have been handled as much as possible in the optimization stage, a final selection step must be taken with the approximated Pareto Front. Here we will emphasize the trade-off visualization.

It is widely accepted that visualization tools are valuable and provide the DM with a meaningful method to analyze the Pareto Front and make decisions [73]. Tools and/or methodologies are required for this final step to successfully embed the DM into the solution refinement and selection process. It is useful if the DM understands and appreciates the impact that a given trade-off in one sub-space could have on others [9]. Even if an EMO process has been applied to a reduced objective space, sometimes the DM needs to increase the space with additional metrics or measurements to have confidence in her/his own decision [9]. Usually, analysis on the Pareto Front may be related with design alternatives comparison and design concepts comparison.

For two-dimensional problems (and sometimes for three-dimensional ones) it is usually straightforward to make an accurate graphical analysis of the Pareto Front (see for example Fig. 2.9), but difficulty increases with the problem dimension. Tools such as VIDEO [68] incorporate a color coding in three-dimensional graphs to analyze trade-offs for 4-dimensional Pareto fronts. In [73], a review on visualization techniques includes techniques such as decision maps, star diagrams, value paths, GAIA, and heatmap graphs. Possibly the most common choices for Pareto Front visualization and analysis in control systems applications are: scatter diagrams, parallel coordinates [60], and level diagrams [8, 109].

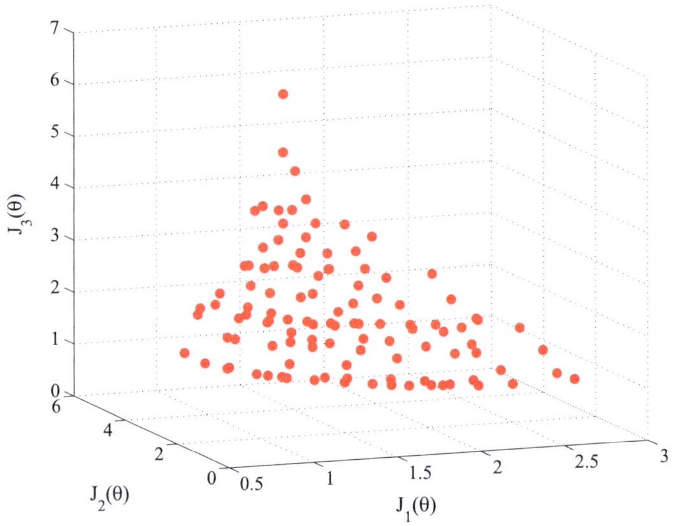

Fig. 2.9 3D Visualization of a 3-dimensional pareto front

Scatter diagram plots (SCp)[8] are straightforward visualizations. They generate an array of 2-D graphs to visualize each combination of a pair of objectives (see Fig. 2.10). This type of visualization is enough for two dimensional problems. To appreciate all the trade-offs of an m-dimensional Pareto Front, at least $\frac{m(m-1)}{2}$ combination plots are required. For example, the Pareto Front of Fig. 2.9 is visualized using SCp in Fig. 2.10. If the DM would like to see the trade-off for an objective and a decision variable from the n-dimensional decision space, she/he will need n times m additional plots.

Parallel coordinate (PAc) visualization strategy [60] plots an m-dimensional objective vector in a two-dimensional graph.[9] For each objective vector $\boldsymbol{J}(\boldsymbol{\theta}) = [J_1(\boldsymbol{\theta}), \ldots, J_m(\boldsymbol{\theta})]$ the ordered pairs $(i, J_i(\boldsymbol{\theta}))$, $i \in [1, \ldots, m]$ are plotted and linked with a line. This is a very compact way of presenting multidimensional information: just one 2-D plot is required. Nevertheless, to entirely represent the trade-off surface some axis relocation may be necessary. For example, in Fig. 2.11, it is possible to appreciate the PAc visualization of the Pareto Front depicted in Fig. 2.9. To appreciate tendencies with the decision space variable, an extended plot with $n+m$ vertical axes is required. An independent graph could be plotted, but some strategy (such as color coding) will be needed to link an objective vector with its corresponding decision vector in order to appreciate trade-off information from the objective space. This kind of feature is incorporated in visualization tools such as TULIP from INRIA,[10]

[8]Tool available in Matlab©.

[9]Tool available in the statistics toolbox of Matlab©.

[10]Available at http://tulip.labri.fr/TulipDrupal/. Includes applications for multidimensional analysis.

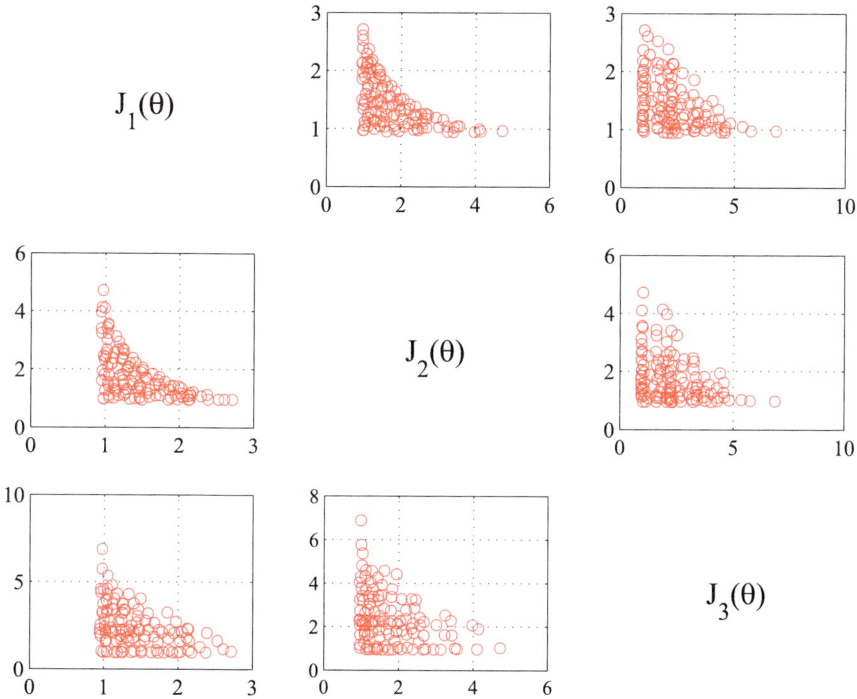

Fig. 2.10 Scatter plot (SCp) visualization for pareto front of Fig. 2.9

which are also helpful for analyzing multidimensional data. Finally, a normalization or y-axis re-scaling can be easily incorporated, if required, to facilitate the analysis.

The *Level Diagrams* (LD) visualization [8][11] is useful for analyzing m-objective Pareto Fronts [145, 146], as it is based on the classification of the approximation $\boldsymbol{J}_P^*$ obtained. Each objective $J_i(\boldsymbol{\theta})$ is normalized $\hat{J}_i(\boldsymbol{\theta})$ with respect to its minimum and maximum values. To each normalized objective vector $\hat{\boldsymbol{J}}(\boldsymbol{\theta})$ a p-norm is applied to evaluate the distance to an ideal[12] solution $\boldsymbol{J}^{ideal}$. The LD tool displays a two dimensional graph for each objective and decision variable. The ordered pairs $\left(J_i(\boldsymbol{\theta}), \|\hat{\boldsymbol{J}}(\boldsymbol{\theta})\|_p\right)$ in each objective sub-graph and $\left(\theta_l, \|\hat{\boldsymbol{J}}(\boldsymbol{\theta})\|_p\right)$ in each decision variable sub-graph are plotted (a total of $n + m$ plots). Therefore, a given solution will have the same y-value in all graphs (see Fig. 2.12). This correspondence will help to evaluate general tendencies along the Pareto Front and to compare solutions according to the selected norm. Also, with this correspondence, information from the objective space is directly embedded in the decision space, since a decision vector inherits its *y-value* from its corresponding objective vector.

[11] GUI for Matlab© is available at: http://www.mathworks.com/matlabcentral/fileexchange/24042.

[12] By default, minimum values for each objective in $\hat{\boldsymbol{J}}(\boldsymbol{\theta})$ could be used to build an ideal solution.

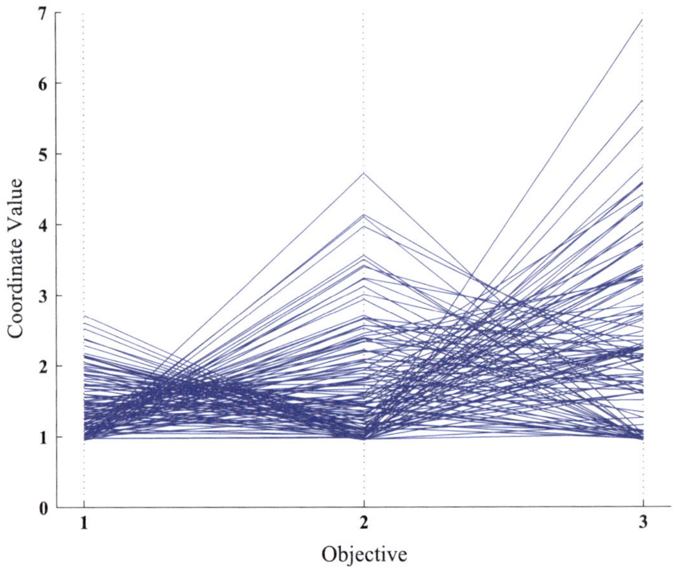

Fig. 2.11 Parallel coordinates plot (PAc) visualization for pareto front of Fig. 2.9

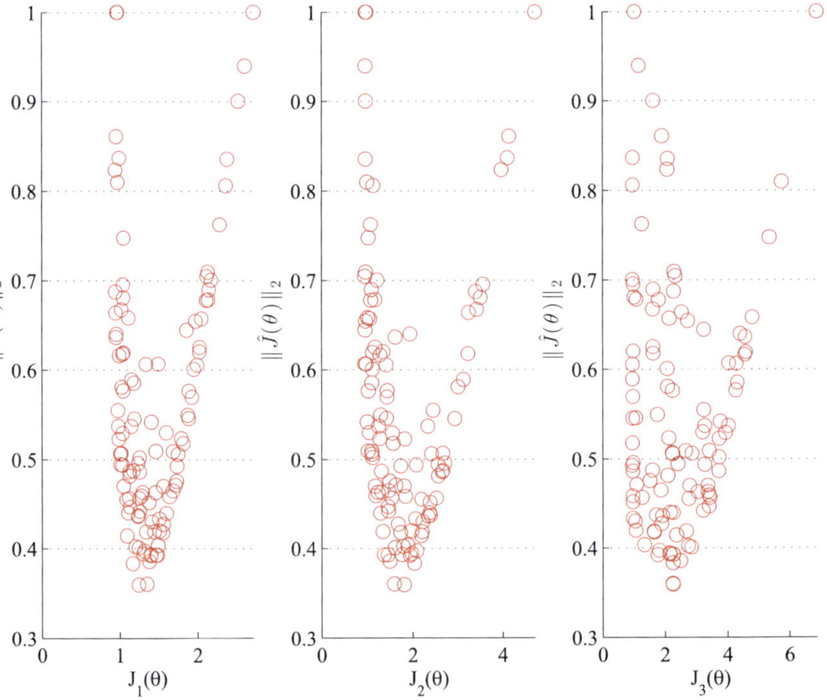

Fig. 2.12 Level diagram (LD) visualization for pareto front of Fig. 2.9

In any case, characteristics required for such a visualization were described in [73]: simplicity (must be understandable); persistence (information must be rememberable by the DM); and completeness (all relevant information must be depicted). Some degree of interactivity with the visualization tool is also desirable (during and/or before the optimization process) to successfully embed the DM into the selection process.

2.3 Related Work in Controller Tuning

As noticed in the previous chapter, multiobjective techniques might be useful for controller tuning applications. This section will provide a brief listing on related work over the last ten years (expanding and updating [115]), with a focus on four controller structures (design concepts): PID-like, State space representation, fuzzy control and model predictive control. While several works have dealt with MOP (using an AOF for example), those where dominance and Pareto Front concepts have been used actively for controller tuning purposes will be included.

Control engineers might select different design objectives in order to evaluate a given controller performance in the feedback loop. According to the basic control loop of Fig. 2.13, such design objectives are typically selected in order to have a measure of:

- Tracking performance of the set point (reference) $r(t)$.
- Rejection performance of load disturbance $d(t)$.
- Robustness to measurement noise $n(t)$.
- Robustness to model uncertainty.

Different measures are used for such purposes, typically in frequency and time domains.

2.3.1 Basic Design Objectives in Frequency Domain

- Maximum value of the complementary sensitivity function.

$$J_{M_p}(\boldsymbol{\theta}) = \left\| P(s)C(s)(I + P(s)C(s))^{-1} \right\|_{\infty} \tag{2.4}$$

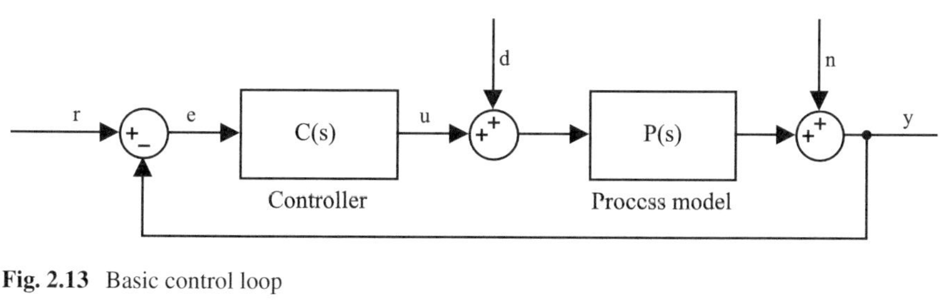

Fig. 2.13 Basic control loop

- Disturbance attenuation performance.

$$J_{W_1}(\boldsymbol{\theta}) = \left\| W(s) \cdot (I + P(s)C(s))^{-1} \right\|_{\infty} < 1 \tag{2.5}$$

- Maximum value of noise sensitivity function.

$$J_{M_u}(\boldsymbol{\theta}) = \left\| C(s)(I + P(s)C(s))^{-1} \right\|_{\infty} \tag{2.6}$$

- Maximum value of the sensitivity function.

$$J_{M_s}(\boldsymbol{\theta}) = \left\| (I + P(s)C(s))^{-1} \right\|_{\infty} \tag{2.7}$$

- Robust stability performance.

$$J_{W_2}(\boldsymbol{\theta}) = \left\| W(s) \cdot (P(s)C(s)(I + P(s)C(s))^{-1}) \right\|_{\infty} < 1 \tag{2.8}$$

where $W(s)$ are weighting transfer functions commonly used in mixed sensitivity techniques.

2.3.2 Basic Design Objectives in Time Domain

- Integral of the absolute error value.

$$J_{IAE}(\boldsymbol{\theta}) = \int_{t=t_0}^{t_f} |r(t) - y(t)| \, dt \tag{2.9}$$

- Integral of the time weighted absolute error value.

$$J_{ITAE}(\boldsymbol{\theta}) = \int_{t=t_0}^{t_f} t \, |r(t) - y(t)| \, dt \tag{2.10}$$

- Integral of the squared error value.

$$J_{ISE}(\boldsymbol{\theta}) = \int_{t=t_0}^{t_f} (r(t) - y(t))^2 \, dt \tag{2.11}$$

- Integral of the time weighted squared error value.

$$J_{ITSE}(\boldsymbol{\theta}) = \int_{t=t_0}^{t_f} t \, (r(t) - y(t))^2 \, dt \tag{2.12}$$

- Settling time: time elapsed from a step change input to the time at which $y(t)$ is within a specified error band of $\Delta\%$.

$$J_{t(100-\Delta)\%}(\boldsymbol{\theta}) \tag{2.13}$$

- Overshoot (for a positive input change).

$$J_{over}(\boldsymbol{\theta}) = \max\left[\max\left(\frac{y(t)-r(t)}{r(t)}\right), 0\right], t \in [t_0, t_f] \tag{2.14}$$

- Maximum deviation (for a load disturbance).

$$J_{overd}(\boldsymbol{\theta}) = \max\left(\left|\frac{y(t)-r(t)}{r(t)}\right|\right), t \in [t_0, t_f] \tag{2.15}$$

- Integral of the squared control action value.

$$J_{ISU}(\boldsymbol{\theta}) = \int_{t=t_0}^{t_f} u(t)^2 dt \tag{2.16}$$

- Integral of the absolute control action value.

$$J_{IAU}(\boldsymbol{\theta}) = \int_{t=t_0}^{t_f} |u(t)| dt \tag{2.17}$$

- Total variation of control action.

$$J_{TV}(\boldsymbol{\theta}) = \int_{t=t_0}^{t_f} \left|\frac{du}{dt}\right| dt \tag{2.18}$$

- Maximum value of control action.

$$J_{maxU}(\boldsymbol{\theta}) = \max(u(t)), t \in [t_0, t_f] \tag{2.19}$$

where $r(t)$, $y(t)$, $u(t)$ are the set-point, controlled variable and manipulated variable respectively in time t. Such objectives, for the sake of simplicity, have been stated in a general sense.

2.3.3 PI-PID Controller Design Concept

PID controllers are reliable control solutions thanks to their simplicity and efficacy [3, 4]. They represent a common solution for industrial applications and therefore, there is still ongoing research on new techniques for robust PID controller tuning [135]. Any improvement in PID tuning is worthwhile, owing to the minimum number of changes required for their incorporation into already operational control loops [125, 130]. As expected, several works have focused on the PID performance improvement.

Given a process model $P(s)$, the following general description for a PID controller of two-degree-of-freedom is used (see Fig. 2.14):

$$\begin{aligned} C(s) = & K_c\left(b + \frac{1}{T_i s^{\lambda}} + c\frac{T_d \cdot s^{\mu}}{\frac{T_d}{N}s^{\mu}+1}\right)R(s) \\ & - K_c\left(1 + \frac{1}{T_i s^{\lambda}} + \frac{T_d \cdot s^{\mu}}{\frac{T_d}{N}s^{\mu}+1}\right)Y(s) \end{aligned} \tag{2.20}$$

where K_c is the proportional gain, T_i the integral time, T_d the derivative time, N the derivative filter, a, b the set-point weighting for proportional and derivative actions; λ and μ are used to represent a PID controller with fractional order [103]. Therefore, the following design concepts (controllers) with their decision variables can be stated:

PI: $\boldsymbol{\theta}_{PI} = [K_c, T_i]$. $b = 1, Td = 0, \lambda = 1$.
PD: $\boldsymbol{\theta}_{PD} = [K_c, T_d]$. $b = c = 1, \frac{1}{N} = 0, \frac{1}{T_i} = 0, \mu = 1$.
PID: $\boldsymbol{\theta}_{PID} = [K_c, T_i, T_d]$. $b = c = 1, \frac{1}{N} = 0, \lambda = 1, \mu = 1$.
PID/N: $\boldsymbol{\theta}_{PID/N} = [K_c, T_i, T_d, N]$. $b = c = \lambda = \mu = 1$.
PI^1: $\boldsymbol{\theta}_{PI^1} = [K_c, T_i, b]$. $T_d = 0, \lambda = 1$.
PID^2: $\boldsymbol{\theta}_{PID^2} = [K_c, T_i, T_d, b, c]$. $\frac{1}{N} = 0, \lambda = \mu = 1$.
PID^2/N: $\boldsymbol{\theta}_{PID^2/N} = [K_c, T_i, T_d, N, b, c]$. , $\lambda = \mu = 1$.
PI$^\lambda$D$^\mu$: $\boldsymbol{\theta}_{FOPID} = [K_c, T_i, T_d, \lambda, \mu]$. $b = c = 1, \frac{1}{N} = 0$.

In Table 2.1 a summary of contributions using these design concepts is provided. Brief remarks on MOP, EMO and MCDM for each work are given. Regarding the MOP, it is important to notice that there are more works focusing on controller tuning for SISO loops; besides, there is also an equilibrium with MOP problems dealing

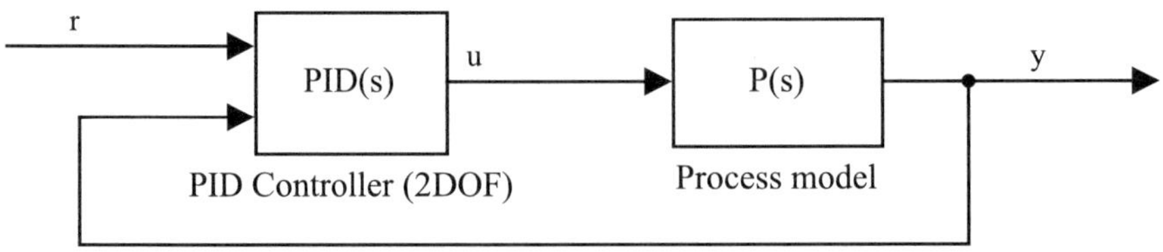

Fig. 2.14 Control loop with a two-degree-of-freedom PID (2DOF-PID) controller

Table 2.1 Summary of MOOD procedures for PID design concept. MOP refers to the number of design objectives; EMO to the algorithm implemented (or used as basis for a new one) in the optimization process. MCDM to the visualization and selection process used

Concept(s)	Process(es)	References	MOP	EMO	MCDM
PID^2/N, PI^1	SISO, MIMO	[53]	4	GA	3D, SCp Concepts comparison
PI^1	FOPDT	[131]	4	GA	3D, 2D Tuning rule methodology
PID	Electromagnetic valve actuator	[126]	7	GA	PAc Iterative controllability analysis for a given design
PID/N	SISO	[57]	3	Ad hoc	SCp Incorporate analysis of time domain objectives
PID	Aeronautical longitudinal control system for an aircraft	[59]	3	SA	3D analysis with other tuning techniques
PD	Mechatronical design (mechanical and control)	[106]	5	GA	SCp design alternatives comparison
PID^2/N	SISO	[108]	15	GA	LD selection according to preferences
PI	Alstom Gasifier MIMO process	[138]	6	NSGA-II	SCp, new indicator included for selection
PID	Flexible AC transmission system	[119]	2	NSGA-II	Fuzzy based selection
PID, I-PD	Chemotherapy control	[1]	2	GA	SCp concepts comparison; intended to support specific treatment
PID	Methanol-Ethanol distillation column, F18/HARV aircraft	[143]	2	PSO	SCp, AFO selection
PI	Wood and Berry MIMO system	[112]	7	DE	LD design alternatives analysis
$PI^{\lambda}D^{\mu}$	SISO	[49]	5	GA	LD design alternatives comparison
$PI^{\lambda}D^{\mu}$	SISO,Hydraulic turbine regulating system	[11]	2	NSGA-II	2D design concepts comparison with a PID
PI, PID/N	Two-area non-reheat thermal system	[102]	3	NSGA-II	Fuzzy-based membership value assignment approach

(continued)

Table 2.1 (continued)

Concept(s)	Process(es)	References	MOP	EMO	MCDM
PI	Speed control reluctance motor	[63]	2	NSGA-II	Selection with an AOF
PID	Twin Rotor MIMO system	[110]	5	DE	LD design alternatives comparison
$PI^{\lambda}D^{\mu}$	Load frequency control	[101]	2	NSGA-II	2D design concepts comparison with PID
$PI^{\lambda}D^{\mu}$	Automatic Voltage Regulator	[100]	2	NSGA-II	2D design concepts comparison
PI	MIMO boiler process	[114]	5	DE	LD design alternatives comparison
PI	MIMO Wood and Berry MIMO system	[111]	7	DE	LD design alternatives comparison
$PI^{\lambda}D^{\mu}$	Automatic Voltage Regulator	[140]	3	Ad hoc	3D design alternative analysis. Design concepts comparison with PID
PI, PID	Two-area non-reheat thermal system; Three-area hydro-thermal power system	[99]	3	GA	SCp, fuzzy-based membership value assignment approach

with 2–3 objectives versus many-objectives optimization. Regarding the optimizer, MOEAs based on GA seem to be more popular for such design concept. In the MCDM stage while a design alternatives comparison is in general performed, the design concepts comparison seems to be more popular when dealing with fractional PID controllers. This is done in order to justify increasing the complexity of the controller. Finally, in the MCDM, classical approaches for visualization based on SCp and 3D representation are the most used, despite the number of objectives managed.

2.3.4 *Fuzzy Controller Design Concept*

Fuzzy systems have been widely and successfully used in control system applications as referenced in [40]. Similar to the use of PID as design concept, the MOOD is useful for analyzing the trade-off between conflicting objectives. In this case, the fuzzy controller is more complex to tune, given its nonlinearity and the major number of variables involved in fuzzyfication, inference and defuzzification steps (see Fig. 2.15).

A comprehensive compendium on the synergy between fuzzy tools and MOEAs is given in [39]. This book will focus on controller implementations. In general, decision variables consider $\boldsymbol{\theta} = [\Lambda, \Upsilon, \overline{\overline{\Lambda}}, \overline{\overline{\Upsilon}}, \mu]$, where:

Λ: is the membership function shape.
$\overline{\overline{\Lambda}}$: is the number of membership functions.
Υ: is the fuzzy rule structure.
$\overline{\overline{\Upsilon}}$: is the number of fuzzy rules.
μ: are the weights of the fuzzy inference system.

In Table 2.2 a summary on these applications is provided. The difference in the quantity of the works dedicated to fuzzy controllers and PID controllers is noticeable.

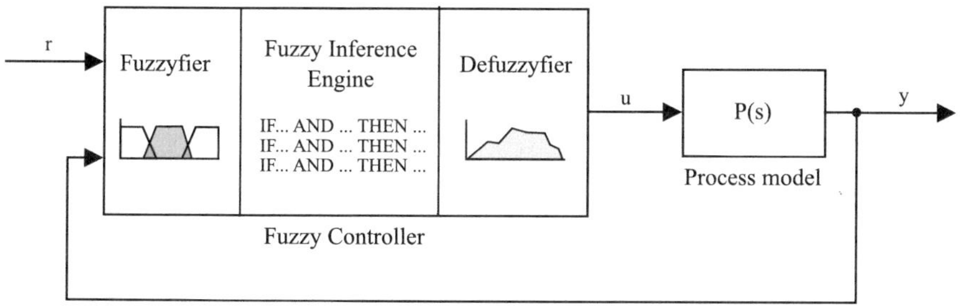

Fig. 2.15 Control loop with a fuzzy controller

Table 2.2 Summary of MOOD procedures for Fuzzy design concept. MOP refers to the number of design objectives; EMO to the algorithm implemented (or used as basis for a new one) in the optimization process. MCDM to the visualization and selection process used

Process(es)	References	MOP	EMO	MCDM
Aeronautical	[12]	9	GA	PAc Constraint violation analysis; fine tuning
DC motor (HiL)	[127]	4	GA	None According performance
Geological	[66]	4	NSGA-II	SCp Design alternatives comparison
Bio-medical	[37]	2	SPEA based	2D Design alternatives/concepts comparison with other controllers. Selection by norm-2 criteria
Mechanical	[80]	3	PSO	3D Design alternatives comparison
HVAC system	[46]	2	SPEA based	2D Design alternatives comparison at two levels: different controllers and different MOEAs
Wall following robot	[56]	4	SPEA based	2D with an AFO

With regard to MOP definition, it seems that EMO has been popular to simultaneously optimize objectives related with performance and interpretation of the fuzzy inference system. Nevertheless, as noticed in [39] scalability issue is a problem worthwhile to address such a design concept. Finally, in the MCDM step, SCp tools have been sufficient for Pareto Front visualization and analysis, due to the low number of objectives stated in the MOP.

2.3.5 State Space Feedback Controller Design Concept

The state space representation has shown to be a remarkable tool for controller design. Several advanced control techniques use this representation to calculate a controller (in the same representation) with a desired performance. In this case, the decision variables are the gains of the matrix K (see Fig. 2.16). Classical optimization approaches in a MOOD framework have been used in [85] with good results. In several instances, it seems that the MOOD procedure has been used to compare classical approaches with the EMO approach, as presented below.

In Table 2.3 a summary on these applications is provided. There are still few works focusing on this design concept.

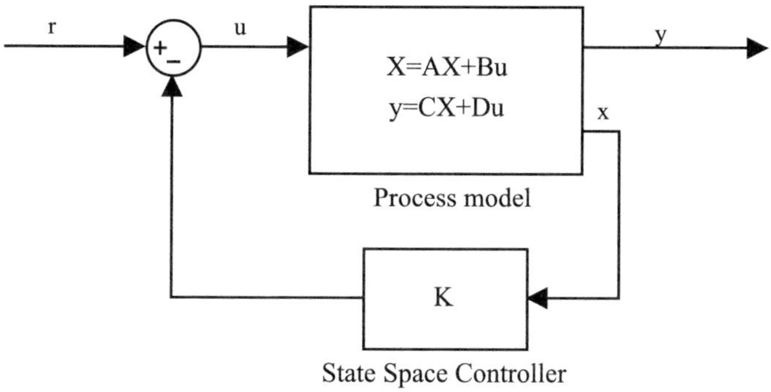

Fig. 2.16 Control loop with a state space controller

Table 2.3 Summary of MOOD procedures for state space representation design concept. MOP refers to the number of design objectives; EMO to the algorithm implemented (or used as basis for a new one) in the optimization process. MCDM to the visualization and selection process used

Process(es)	References	MOP	EMO	MCDM
SISO, MIMO	[54]	3	GA	SCp Concepts comparison with LMI design
SISO	[94]	3	GA	2D Concepts comparison with LMI
Mechanical	[62]	4	GA	SCp Design alternatives comparison
Networked Predictive control, various examples	[25]	2	NSGA-II with LMIs	2D Design alternatives analysis on examples
Biped robot	[76]	2	MOPSO and NSGA-II	2D Design alternatives analysis on examples
Twin Rotor MIMO system	[110]	18	DE	LD Design concepts comparison with a PID controller; design alternatives comparison

2.3.6 Predictive Control Design Concept

On-line applications for MOOD are not straightforward, since the MCDM stage must be carried out, in some instances, automatically. As a result, analysis that relies on the DM must be codified to become an automatic process. Approaches using EMO in the MOOD procedure are presented below; where decision variables $\boldsymbol{\theta}$ is conformed by the control action $\boldsymbol{u}$ through the control horizon, see Fig. 2.17.

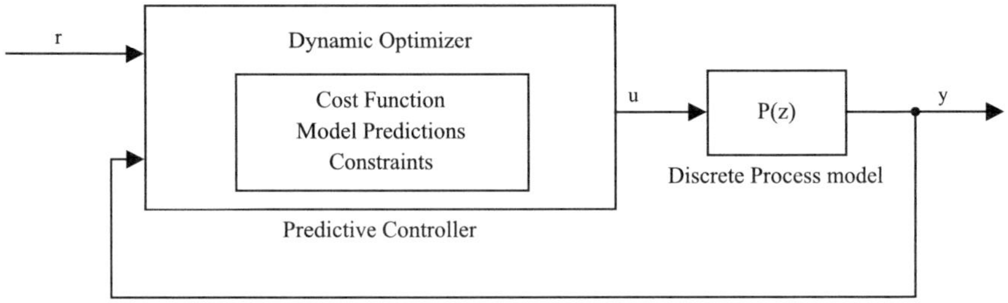

Fig. 2.17 Control loop with a predictive controller

Table 2.4 Summary of MOOD procedures for predictive control design concept. MOP refers to the number of design objectives; EMO to the algorithm implemented (or used as basis for a new one) in the optimization process. MCDM to the visualization and selection process used

Process(es)	References	MOP	EMO	MCDM
Mechanical	[47]	2	GA	Fuzzy inference system is used
Chemical	[13]	8	NSGA-II	Successive ordering according to feasibility
Subway ventilation system	[72]	2	NSGA-II	Decision rule
Smart energy efficient buildings	[118]	2	GA	Decision rule

In Table 2.4 a summary on these applications is provided. Predictive control seems to be an opportunity to apply the MOOD approach, due to the few works dedicated to this control design alternative. Nevertheless, it can also be seen that the problem relies on tracking the Pareto Front each sampling time.

2.4 Conclusions on This Chapter

In this chapter, fundamental concepts regarding multiobjective optimization were introduced. Besides, some notions and remarks on the fundamental steps of a holistic MOOD procedure were commented: the MOP definition, the EMO process and the MCDM stage.

Furthermore, related work on controller tuning applications using such MOOD techniques was revisited including works were dominance and Pareto Front concepts were used actively for controller tuning purposes. Design concepts (controller structures) listed were PID-like controllers, fuzzy structures, state space representation and model predictive control. Even focusing on contributions using EMO, there are also examples solving MOPs with other (deterministic) techniques, for example:

- PID-like: [71, 121].
- State space representation: [137].
- Predictive control: [6, 97, 105, 133].
- Optimal control: [132, 134].

As commented in the previous chapter, MOOD procedures might be a useful tool for controller tuning purposes. With such techniques, it is possible appreciating trade-off between conflictive control objectives (performance and robustness for instance). Which is important to remember is the fundamental question for such techniques:

- *What kind of problems are worth to address with MOOD?*

That question leads to others:

- *Is it difficult to find a controller with a reasonable trade-off among design objectives?*
- *Is it worthwhile analysing the trade-off among controllers (design alternatives)?*

If the answer is yes to both questions, then the MOOD procedure could be an appropriate tool for the problem at hand. Otherwise, other tuning techniques or AOF approaches could be enough.

During the remaining chapters a set of tools and algorithms for EMO and MCDM stage will be presented, in order to provide to the readers an introductory toolbox for MOOD procedures.

References

1. Algoul S, Alam M, Hossain M, Majumder M (2011) Multi-objective optimal chemotherapy control model for cancer treatment. Med Biol Eng Comput 49:51–65. doi:10.1007/s11517-010-0678-y
2. Aslam T, Ng A (2010) Multi-objective optimization for supply chain management: a literature review and new development. In: 2010 8th international conference on supply chain management and information systems (SCMIS) (Oct 2010), pp 1 –8
3. Åström K, Hägglund T (2001) The future of PID control. Control Eng Pract 9(11):1163–1175
4. Åström KJ, Hägglund T (2005) Advanced PID Control. ISA Instrum Syst Autom Soc Res Triangle Park, NC 27709
5. Batista L, Campelo F, Guimar£es F, Ramrez J (2011) Pareto cone ϵ-dominance: improving convergence and diversity in multiobjective evolutionary algorithms. In: Takahashi R, Deb K, Wanner E, Greco S (eds) Evolutionary multi-criterion optimization vol. 6576 of Lecture notes in computer science. Springer, Heidelberg, pp 76–90. doi:10.1007/978-3-642-19893-9-6
6. Bemporad A, noz de la Peña DM (2009) Multiobjective model predictive control. Automatica 45(12):2823–2830
7. Beyer H-G, Sendhoff B (2007) Robust optimization - a comprehensive survey. Comput Meth Appl Mech Eng 196(33–34):3190–3218

8. Blasco X, Herrero J, Sanchis J, Martínez M (2008) A new graphical visualization of n-dimensional pareto front for decision-making in multiobjective optimization. Inf Sci 178(20):3908–3924
9. Bonissone P, Subbu R, Lizzi J (2009) Multicriteria decision making (MCDM): a framework for research and applications. IEEE Comput Intell Mag 4(3):48 –61 (2009)
10. Branke J, Schmeck H, Deb K, Reddy SM (2004) Parallelizing multi-objective evolutionary algorithms: cone separation. In: Congress on evolutionary computation, 2004. CEC2004 (June 2004), vol 2, pp 1952–1957
11. Chen Z, Yuan X, Ji B, Wang P, Tian H (2014) Design of a fractional order PID controller for hydraulic turbine regulating system using chaotic non-dominated sorting genetic algorithm II. Energy Convers Manag 84:390–404
12. Chipperfield A, Bica B, Fleming P (2002) Fuzzy scheduling control of a gas turbine aero-engine: a multiobjective approach. IEEE Trans Indus Electron 49(3):536–548
13. Chuk OD, Kuchen BR (2011) Supervisory control of flotation columns using multi-objective optimization. Miner Eng 24(14):1545–1555
14. Coello C (2000) Handling preferences in evolutionary multiobjective optimization: a survey. In: Proceedings of the 2000 congress on evolutionary computation, vol 1, pp 30–37
15. Coello C (2011) An introduction to multi-objective particle swarm optimizers. In: Gaspar-Cunha A, Takahashi R, Schaefer G, Costa L (eds) Soft computing in industrial applications, vol 96 of Advances in intelligent and soft computing. Springer, Heidelberg, pp 3–12. doi:10.1007/978-3-642-20505-7_1
16. Coello CAC (2002) Theorical and numerical constraint-handling techniques used with evolutionary algorithms: a survey of the state of the art. Comput Meth Appl Mech Eng 191:1245–1287
17. Coello CAC, Lamont GB (2004) Applications of multi-objective evolutionary algorithms. In: Advances in natural computation, vol 1. World Scientific Publishing
18. Coello CAC, Lamont GB, Veldhuizen DAV (2007) Multi-criteria decision making. In: Evolutionary algorithms for solving multi-objective problems. Genetic and evolutionary computation series. Springer US, pp 515–545
19. Coello CAC., Veldhuizen DV, Lamont G (2002) Evolutionary algorithms for solving multi-objective problems. Kluwer Academic Press
20. Coello Coello C (2006) Evolutionary multi-objective optimization: a historical view of the field. IEEE Comput Intellig Magaz 1(1):28–36
21. Coello Coello C (2011) Evolutionary multi-objective optimization: basic concepts and some applications in pattern recognition. In: Martnez-Trinidad J, Carrasco-Ochoa J, Ben-Youssef Brants C, Hancock E (eds.) Pattern recognition, vol 6718 of Lecture notes in computer science. Springer, Heidelberg, pp 22–33. doi:10.1007/978-3-642-21587-2_3
22. Corne DW, Knowles JD (2007) Techniques for highly multiobjective optimisation: some nondominated points are better than others. In: Proceedings of the 9th annual conference on genetic and evolutionary computation (New York, NY, USA, 2007), GECCO '07, ACM, pp 773–780
23. Cruz C, González JR, Pelta DA (2011) Optimization in dynamic environments: a survey on problems, methods and measures. Soft Comput 15:1427–1448
24. Das I, Dennis J (1998) Normal-boundary intersection: a new method for generating the pareto surface in non-linear multicriteria optimization problems. SIAM J Optim 8:631–657
25. Das S, Das S, Pan I (2013) Multi-objective optimization framework for networked predictive controller design. ISA Trans 52(1):56–77
26. Das S, Maity S, Qu B-Y, Suganthan P (2011) Real-parameter evolutionary multimodal optimization - a survey of the state-of-the-art. Swarm Evol Comput 1(2):71–88
27. Das S, Mullick SS, Suganthan P (2016) Recent advances in differential evolution an updated survey. Swarm Evol Comput 27:1–30
28. Das S, Suganthan PN (2010) Differential evolution: a survey of the state-of-the-art. IEEE Trans Evol Comput 99:1–28

29. Deb K (2000) An efficient constraint handling method for genetic algorithms. Comput Meth Appl Mech Eng 186(2–4):311–338
30. Deb K (2012) Advances in evolutionary multi-objective optimization. In: Fraser G, Teixeira de Souza J (eds) Search based software engineering, vol 7515 of Lecture notes in computer science. Springer, Berlin, Heidelberg, pp 1–26
31. Deb K, Pratap A, Agarwal S, Meyarivan T (2002) A fast and elitist multiobjective genetic algorithm: NSGA-II. IEEE Trans Evol Comput 6(2):124–141
32. Deb K, Saha A (2002) Multimodal optimization using a bi-objective evolutionary algorithm. Evol Comput 27–62
33. Dorigo M, Sttzle T (2010) Ant colony optimization: overview and recent advances. In: Gendreau M, Potvin J-Y (eds) Handbook of metaheuristics, vol 146 of International series in operations research & management science. Springer US, pp 227–263
34. Efstratiadis A, Koutsoyiannis D (2010) One decade of multi-objective calibration approaches in hydrological modelling: a review. Hydrol Sci J 55(1):58–78
35. Fadaee M, Radzi M (2012) Multi-objective optimization of a stand-alone hybrid renewable energy system by using evolutionary algorithms: a review. Renew Sustain Energy Rev 16(5):3364–3369
36. Farina M, Deb K, Amato P (2004) Dynamic multiobjective optimization problems: test cases, approximations, and applications. IEEE Trans Evol Comput 8(5):425–442
37. Fazendeiro P, de Oliveira J, Pedrycz W (2007) A multiobjective design of a patient and anaesthetist-friendly neuromuscular blockade controller. IEEE Trans Biomed Eng 54(9):1667–1678
38. Fazlollahi S, Mandel P, Becker G, Maréchal F (2012) Methods for multi-objective investment and operating optimization of complex energy systems. Energy 45(1):12–22
39. Fazzolar M, Alcalá R, Nojima Y, Ishibuchi H, Herrera F (2013) A review of the application of multi-objective evolutionary fuzzy systems: current status and further directions. IEEE Trans Fuzzy Syst 21(1):45–65
40. Feng G (2006) A survey on analysis and design of model-based fuzzy control systems. IEEE Trans Fuzzy Syst 14(5):676–697
41. Figueira J, Greco S, Ehrgott M (2005) Multiple criteria decision analysis: state of the art surveys. Springer International Series
42. Fister I Jr, Yang, X-S, Brest J (2013) A comprehensive review of firefly algorithms. Swarm Evol Comput 13:34–46
43. Fleming P, Purshouse R (2002) Evolutionary algorithms in control systems engineering: a survey. Control Eng Pract 10:1223–1241
44. Fonseca C, Fleming P (1998) Multiobjective optimization and multiple constraint handling with evolutionary algorithms-I: a unified formulation. IEEE Trans Systems, Man Cybern Part A: Syst Humans 28(1):26–37
45. Fonseca C, Fleming P (1998) Multiobjective optimization and multiple constraint handling with evolutionary algorithms-II: application example. IEEE Trans Systems, Man Cybern Part A: Syst Humans 28(1):38–47
46. Gacto M, Alcalá R, Herrera F (2012) A multi-objective evolutionary algorithm for an effective tuning of fuzzy logic controllers in heating, ventilating and air conditioning systems. Appl Intell 36:330–347. doi:10.1007/s10489-010-0264-x
47. Garca JJV, Garay VG, Gordo EI, Fano FA, Sukia ML (2012) Intelligent multi-objective nonlinear model predictive control (iMO-NMPC): towards the on-line optimization of highly complex control problems. Expert Syst Appl 39(7):6527–6540
48. Gong W, Cai Z, Zhu L (2009) An efficient multiobjective. Differential Evolution algorithm for engineering design. Struct Multidisciplinary Optim 38:137–157. doi:10.1007/s00158-008-0269-9
49. Hajiloo A, Nariman-zadeh N, Moeini A (2012) Pareto optimal robust design of fractional-order PID controllers for systems with probabilistic uncertainties. Mechatronics 22(6):788–801

50. Harik G, Lobo F, Goldberg D (1999) The compact genetic algorithm. IEEE Trans Evol Comput 3(4):287–297
51. Hernández-Daz AG, Santana-Quintero LV, Coello CAC, Molina J (2007) Pareto-adaptive ϵ-dominance. Evol Comput 4:493–517
52. Herrero J, Martínez M, Sanchis J, Blasco X (2007) Well-distributed Pareto front by using the ϵ-MOGA evolutionary algorithm. In: Computational and ambient intelligence, vol LNCS 4507. Springer-Verlag, pp 292–299
53. Herreros A, Baeyens E, Perán JR (2002) Design of PID-type controllers using multiobjective genetic algorithms. ISA Trans 41(4):457–472
54. Herreros A, Baeyens E, Perán JR (2002) MRCD: a genetic algorithm for multiobjective robust control design. Eng Appl Artif Intell 15:285–301
55. Houska B, Ferreau HJ, Diehl M (2011) ACADO toolkit an open source framework for automatic control and dynamic optimization. Optim Control Appl Meth 32(3):298–312
56. Hsu C-H, Juang C-F (2013) Multi-objective continuous-ant-colony-optimized fc for robot wall-following control. Comput Intell Mag IEEE 8(3):28–40
57. Huang L, Wang N, Zhao J-H (2008) Multiobjective optimization for controller design. Acta Automatica Sinica 34(4):472–477
58. Huang V, Qin A, Deb K, Zitzler E, Suganthan P, Liang J, Preuss M, Huband S (2007) Problem definitions for performance assessment on multi-objective optimization algorithms. Nanyang Technological University, Singapore, Tech. rep
59. Hung M-H, Shu L-S, Ho S-J, Hwang S-F, Ho S-Y (2008) A novel intelligent multiobjective simulated annealing algorithm for designing robust PID controllers. IEEE Trans Syst Man Cybern Part A: Syst Humans 38(2):319–330
60. Inselberg A (1985) The plane with parallel coordinates. Visual Comput 1:69–91
61. Ishibuchi H, Tsukamoto N, Nojima Y (2008) Evolutionary many-objective optimization: a short review. In: CEC 2008. (IEEE World Congress on Computational Intelligence). IEEE Congress on Evolutionary Computation, 2008 (June 2008), pp 2419–2426
62. Jamali A, Hajiloo A, Nariman-zadeh N (2010) Reliability-based robust pareto design of linear state feedback controllers using a multi-objective uniform-diversity genetic algorithm (MUGA). Expert Syst Appl 37(1):401–413
63. Kalaivani L, Subburaj P, Iruthayarajan MW (2013) Speed control of switched reluctance motor with torque ripple reduction using non-dominated sorting genetic algorithm (nsga-ii). Int J Electr Power Energy Syst 53:69–77
64. Karaboga D, Gorkemli B, Ozturk C, Karaboga N (2012) A comprehensive survey: artificial bee colony (ABC) algorithm and applications. Artif Intell Rev 1–37
65. Kennedy J, Eberhart R (1995) Particle swarm optimization. In: Proceedings IEEE International Conference on Neural Networks, vol 4, pp 1942–1948
66. Kim H-S, Roschke PN (2006) Fuzzy control of base-isolation system using multi-objective genetic algorithm. Comput-Aided Civil Infrastruct Eng 21(6):436–449
67. Knowles J, Thiele L, Zitzler E (2014) A tutorial on the performance assessment of stochastic multiobjective optimizers. Tech. Rep. TIK report No. 214, Computer Engineering and networks laboratory. ETH Zurich, 2006
68. Kollat JB, Reed P (2007) A framework for visually interactive decision-making and design using evolutionary multi-objective optimization (VIDEO). Environ Modell Softw 22(12):1691–1704
69. Konak A, Coit DW, Smith AE (2006) Multi-objective optimization using genetic algorithms: a tutorial. Reliab Eng Syst Safety 91(9):992–1007. Special Issue - Genetic Algorithms and Reliability
70. Laumanns M, Thiele L, Deb K, Zitzler E (2002) Combining convergence and diversity in evolutionary multiobjective optimization. Evol Comput 3:263–282
71. Leiva MC, Rojas JD (2015) New tuning method for pi controllers based on pareto-optimal criterion with robustness constraint. IEEE Latin America Trans 13(2):434–440
72. Liu H, Lee S, Kim M, Shi H, Kim JT, Wasewar KL, Yoo C (2013) Multi-objective optimization of indoor air quality control and energy consumption minimization in a subway ventilation system. Energy Build 66:553–561

73. Lotov A, Miettinen K (2008) Visualizing the pareto frontier. In: Branke J, Deb K, Miettinen K, Slowinski R (eds) Multiobjective optimization, vol 5252 of Lecture notes in computer science. Springer, Heidelberg, pp 213–243
74. Lozano M, Molina D, Herrera F (2011) Soft computing: special issue on scalability of evolutionary algorithms and other metaheuristics for large-scale continuous optimization problems, vol 15. Springer-Verlag
75. Lygoe R, Cary M, Fleming P (2010) A many-objective optimisation decision-making process applied to automotive diesel engine calibration. In: Deb K, Bhattacharya A, Chakraborti N, Chakroborty P, Das S, Dutta J, Gupta S, Jain A, Aggarwal V, Branke J, Louis S, Tan K (eds) Simulated evolution and learning, vol 6457 of Lecture notes in computer science. Springer, Heidelberg, pp 638–646. doi:10.1007/978-3-642-17298-4_72
76. Mahmoodabadi M, Taherkhorsandi M, Bagheri A (2014) Pareto design of state feedback tracking control of a biped robot via multiobjective pso in comparison with sigma method and genetic algorithms: Modified nsgaii and matlabs toolbox. Scientific World J
77. Mallipeddi R, Suganthan P (2009) Problem definitions and evaluation criteria for the CEC 2010 competition on constrained real-parameter optimization. Nanyang Technological University, Singapore, Tech. rep
78. Mallipeddi R, Suganthan P (2010) Ensemble of constraint handling techniques. IEEE Trans Evol Comput 14(4):561–579
79. Mansouri SA, Gallear D, Askariazad MH (2012) Decision support for build-to-order supply chain management through multiobjective optimization. Int J Prod Econ 135(1):24–36
80. Marinaki M, Marinakis Y, Stavroulakis G (2011) Fuzzy control optimized by a multi-objective particle swarm optimization algorithm for vibration suppression of smart structures. Struct Multidisciplinary Optim 43:29–42. doi:10.1007/s00158-010-0552-4
81. Marler R, Arora J (2004) Survey of multi-objective optimization methods for engineering. Struct Multidisciplinary Optim 26:369–395
82. Martínez M, Herrero J, Sanchis J, Blasco X, García-Nieto S (2009) Applied Pareto multi-objective optimization by stochastic solvers. Eng Appl Artif Intell 22:455–465
83. Martins JRRA, Lambe AB (2013) Multidisciplinary design optimization: a survey of architectures. AIAA J 51(9):2049–2075
84. Mattson CA, Messac A (2005) Pareto frontier based concept selection under uncertainty, with visualization. Optim Eng 6:85–115
85. Meeuse F, Tousain RL (2002) Closed-loop controllability analysis of process designs: application to distillation column design. Comput Chem Eng 26(4–5):641–647
86. Messac A, Ismail-Yahaya A, Mattson C (2003) The normalized normal constraint method for generating the pareto frontier. Struct Multidisciplinary Optim 25:86–98
87. Messac A, Mattson C (2002) Generating well-distributed sets of pareto points for engineering design using physical programming. Optim Eng 3:431–450. doi:10.1023/A:1021179727569
88. Metaxiotis K, Liagkouras K (2012) Multiobjective evolutionary algorithms for portfolio management: a comprehensive literature review. Expert Syst Appl 39(14):11685–11698
89. Mezura-Montes E, Coello CAC (2011) Constraint-handling in nature-inspired numerical optimization: past, present and future. Swarm Evol Comput 1(4):173–194
90. Mezura-Montes E, Reyes-Sierra M, Coello C (2008) Multi-objective optimization using differential evolution: a survey of the state-of-the-art. Adv Differ Evol SCI 143:173–196
91. Miettinen KM (1998) Nonlinear multiobjective optimization. Kluwer Academic Publishers
92. Mininno E, Neri F, Cupertino F, Naso D (2011) Compact differential evolution. IEEE Trans Evol Comput 15(1):32–54
93. Mohan BC, Baskaran R (2012) A survey: ant colony optimization based recent research and implementation on several engineering domain. Expert Syst Appl 39(4):4618–4627
94. Molina-Cristóbal A, Griffin I, Fleming P, Owens D (2006) Linear matrix inequialities and evolutionary optimization in multiobjective control. Int J Syst Sci 37(8):513–522
95. Moscato P, Cotta C (2010) A modern introduction to memetic algorithms. In: Gendreau M, Potvin J-Y (eds) Handbook of metaheuristics, vol 146 International series in operations research & management science. Springer US, pp 141–183

96. Munro M, Aouni B (2012) Group decision makers' preferences modelling within the goal programming model: an overview and a typology. J Multi-Criteria Dec Anal 19(3–4):169–184
97. MZavala V, Flores-Tlacuahuac A (2012) Stability of multiobjective predictive control: an utopia-tracking approach. Automatica 48(10):2627–2632
98. Neri F, Cotta C (2012) Memetic algorithms and memetic computing optimization: a literature review. Swarm Evol Comput 2:1–14
99. Nikmanesh E, Hariri O, Shams H, Fasihozaman M (2016) Pareto design of load frequency control for interconnected power systems based on multi-objective uniform diversity genetic algorithm (muga). Int J Electric Power Energy Syst 80:333–346
100. Pan I, Das S (2013) Frequency domain design of fractional order PID controller for AVR system using chaotic multi-objective optimization. Int J Electric Power Energy Syst 51:106–118
101. Pan I, Das S (2015) Fractional-order load-frequency control of interconnected power systems using chaotic multi-objective optimization. Appl Soft Comput 29:328–344
102. Panda S, Yegireddy NK (2013) Automatic generation control of multi-area power system using multi-objective non-dominated sorting genetic algorithm-ii. Int J Electric Power Energy Syst 53:54–63
103. Podlubny I (1999) Fractional-order systems and pi/sup /spl lambda//d/sup /spl mu//-controllers. IEEE Trans Autom Control 44(1):208–214
104. Purshouse R, Fleming P (2007) On the evolutionary optimization of many conflicting objectives. IEEE Trans Evol Comput 11(6):770–784
105. Ramrez-Arias A, Rodrguez F, Guzmán J, Berenguel M (2012) Multiobjective hierarchical control architecture for greenhouse crop growth. Automatica 48(3):490–498
106. Rao JS, Tiwari R (2009) Design optimization of double-acting hybrid magnetic thrust bearings with control integration using multi-objective evolutionary algorithms. Mechatronics 19(6):945–964
107. Reed P, Hadka D, Herman J, Kasprzyk J, Kollat J (2013) Evolutionary multiobjective optimization in water resources: the past, present, and future. Adv Water Res 51(1):438–456
108. Reynoso-Meza G, Blasco X, Sanchis J (2009) Multi-objective design of PID controllers for the control benchmark 2008–2009 (in spanish). Revista Iberoamericana de Automática e Informática Industrial 6(4):93–103
109. Reynoso-Meza G, Blasco X, Sanchis J, Herrero JM (2013) Comparison of design concepts in multi-criteria decision-making using level diagrams. Inf Sci 221:124–141
110. Reynoso-Meza G, García-Nieto S, Sanchis J, Blasco X (2013) Controller tuning using multiobjective optimization algorithms: a global tuning framework. IEEE Trans Control Syst Technol 21(2):445–458
111. Reynoso-Meza G, Sanchis J, Blasco X, Freire RZ (2016) Evolutionary multi-objective optimisation with preferences for multivariable PI controller tuning. Expert Syst Appl 51:120–133
112. Reynoso-Meza G, Sanchis J, Blasco X, Herrero JM (2012) Multiobjective evolutionary algortihms for multivariable PI controller tuning. Expert Syst Appl 39:7895–7907
113. Reynoso-Meza G, Sanchis J, Blasco X, Martínez M (2010) Multiobjective design of continuous controllers using differential evolution and spherical pruning. In: Chio CD, Cagnoni S, Cotta C, Eber M, Ekárt A, Esparcia-Alcaráz AI, Goh CK, Merelo J, Neri F, Preuss M, Togelius J, Yannakakis GN (eds) Applications of evolutionary computation, Part I (2010) vol LNCS 6024, Springer-Verlag, pp 532–541
114. Reynoso-Meza G, Sanchis J, Blasco X, Martínez M (2016) Preference driven multi-objective optimization design procedure for industrial controller tuning. Inf Sci 339:108–131
115. Reynoso-Meza G, Sanchis J, Blasco X, Martínez M Controller tuning using evolutionary multi-objective optimisation: current trends and applications. Control Eng Pract (20XX) (Under revision)
116. Sanchis J, Martínez M, Blasco X, Salcedo JV (2008) A new perspective on multiobjective optimization by enhanced normalized normal constraint method. Struct Multidisciplinary Optim 36:537–546

117. Santana-Quintero L, Montaño A, Coello C (2010) A review of techniques for handling expensive functions in evolutionary multi-objective optimization. In: Tenne Y, Goh C-K (eds) Computational intelligence in expensive optimization problems, vol 2 of Adaptation learning and optimization. Springer, Heidelberg, pp 29–59
118. Shaikh PH, Nor NBM, Nallagownden P, Elamvazuthi I, Ibrahim T (2016) Intelligent multi-objective control and management for smart energy efficient buildings. Int J Electric Power Energy Syst 74:403–409
119. Sidhartha Panda (2011) Multi-objective PID controller tuning for a facts-based damping stabilizer using non-dominated sorting genetic algorithm-II. Int J Electr Power Energy Syst 33(7):1296–1308
120. Singh H, Isaacs A, Ray T (2011) A Pareto corner search evolutionary algorithm and dimensionality reduction in many-objective optimization problems. IEEE Trans Evol Comput 15(4):539–556
121. Snchez HS, Vilanova R (2013) Multiobjective tuning of pi controller using the nnc method: simplified problem definition and guidelines for decision making. In: 2013 IEEE 18th conference on emerging technologies factory automation (ETFA) (Sept 2013), pp 1–8
122. Srinivas M, Patnaik L (1994) Genetic algorithms: a survey. Computer 27(6):17–26
123. Srinivasan S, Ramakrishnan S (2011) Evolutionary multi objective optimization for rule mining: a review. Artif Intell Rev 36:205–248. doi:10.1007/s10462-011-9212-3
124. Stengel RF, Marrison CI (1992) Robustness of solutions to a benchmark control problem. J Guid Control Dyn 15:1060–1067
125. Stewart G, Samad T (2011) Cross-application perspectives: application and market requirements. In: Samad T, Annaswamy A (eds) The impact of control technology. IEEE Control Systems Society, pp 95–100
126. Stewart P, Gladwin D, Fleming P (2007) Multiobjective analysis for the design and control of an electromagnetic valve actuator. Proc Inst Mech Eng Part D: J Autom Eng 221:567–577
127. Stewart P, Stone D, Fleming P (2004) Design of robust fuzzy-logic control systems by multi-objective evolutionary methods with hardware in the loop. Eng Appl Artif Intell 17(3): 275–284
128. Storn R, Price K (1997) Differential evolution: a simple and efficient heuristic for global optimization over continuous spaces. J Global Optim 11:341–359
129. Sun Y, Zhang C, Gao L, Wang X (2011) Multi-objective optimization algorithms for flow shop scheduling problem: a review and prospects. Int J Adv Manuf Technol 55:723–739. doi:10.1007/s00170-010-3094-4
130. Tan W, Liu J, Fang F, Chen Y (2004) Tuning of PID controllers for boiler-turbine units. ISA Trans 43(4):571–583
131. Tavakoli S, Griffin I, Fleming P (2007) Multi-objective optimization approach to the PI tuning problem. In: Proceedings of the IEEE congress on evolutionary computation (CEC2007), pp 3165–3171
132. Vallerio M, Hufkens J, Impe JV, Logist F (2015) An interactive decision-support system for multi-objective optimization of nonlinear dynamic processes with uncertainty. Expert Syst Appl 42(21):7710–7731
133. Vallerio M, Impe JV, Logist F (2014) Tuning of NMPC controllers via multi-objective optimisation. Comput Chem Eng 61:38–50
134. Vallerio M, Vercammen D, Impe JV, Logist F (2015) Interactive NBI and (e)nnc methods for the progressive exploration of the criteria space in multi-objective optimization and optimal control. Comput Chem Eng 82:186–201
135. Vilanova R, Alfaro VM (2011) Robust PID control: an overview (in spanish). Revista Iberoamericana de Automática e Informática Industrial 8(3):141–158
136. Wie B, Bernstein DS (1992) Benchmark problems for robust control design. J Guidance Control Dyn 15:1057–1059
137. Xiong F-R, Qin Z-C, Xue Y, Schtze O, Dıng Q, Sun J-Q (2014) Multi-objective optimal design of feedback controls for dynamical systems with hybrid simple cell mapping algorithm. Commun Nonlinear Sci Numer Simul 19(5):1465–1473

138. Xue Y, Li D, Gao F (2010) Multi-objective optimization and selection for the PI control of ALSTOM gasifier problem. Control Eng Pract 18(1):67–76
139. Yusup N, Zain AM, Hashim SZM (2012) Evolutionary techniques in optimizing machining parameters: review and recent applications (2007–2011). Expert Syst Appl 39(10):9909–9927
140. Zeng G-Q, Chen J, Dai Y-X, Li L-M, Zheng C-W, Chen M-R (2015) Design of fractional order PID controller for automatic regulator voltage system based on multi-objective extremal optimization. Neurocomputing 160:173–184
141. Zhang Q, Li H (2007) MOEA/D: a multiobjective evolutionary algorithm based on decomposition. IEEE Trans Evol Comput 11(6):712–731
142. Zhang Q, Zhou A, Zhao S, Suganthan P, Liu W, Tiwari S (2008) Multiobjective optimization test instances for the cec 2009 special session and competition. Tech. Rep. CES-887, University of Essex and Nanyang Technological University
143. Zhao S-Z, Iruthayarajan MW, Baskar S, Suganthan P (2011) Multi-objective robust PID controller tuning using two lbests multi-objective particle swarm optimization. Inf Sci 181(16):3323–3335
144. Zhou A, Qu B-Y, Li H, Zhao S-Z, Suganthan PN, Zhang Q (2011) Multiobjective evolutionary algorithms: a survey of the state of the art. Swarm Evol Comput 1(1):32–49
145. Zio E, Bazzo R (2011) Level diagrams analysis of pareto front for multiobjective system redundancy allocation. Reliab Eng Syst Safety 96(5):569–580
146. Zio E, Razzo R (2010) Multiobjective optimization of the inspection intervals of a nuclear safety system: a clustering-based framework for reducing the pareto front. Ann Nuclear Energy 37:798–812
147. Zitzler E, Knzli S (2004) Indicator-based selection in multiobjective search. In Yao X, Burke E, Lozano J, Smith J, Merelo-Guervós J, Bullinaria J, Rowe J, Tino P, Kabán A, Schwefel H-P (eds) Parallel problem solving from nature - PPSN VIII, vol 3242 of Lecture notes in computer science. Springer, Heidelberg, pp 832–842. doi:10.1007/978-3-540-30217-9_84
148. Zitzler E, Thiele L, Laumanns M, Fonseca C, da Fonseca V (2003) Performance assessment of multiobjective optimizers: an analysis and review. IEEE Trans Evol Comput 7(2):117–132

Chapter 3
Tools for the Multiobjective Optimization Design Procedure

Abstract In this chapter, tools for the evolutionary multiobjective optimization process and the multicriteria decision making stage to be used throughout this book (as a reference) are presented. Regarding the optimization process, three different versions of a multiobjective evolutionary algorithm based on Differential Evolution will be commented; with those proposals, features such as convergence, diversity and pertinency are considered. Regarding the decision making stage, Level Diagrams will be introduced, due to their capabilities to analyze m-dimensional Pareto fronts.

3.1 EMO Process

In this section, we will focus on the second stage of the MOOD procedure: the multiobjective optimization process (Fig. 3.1). In the previous chapter, desirable characteristics for multiobjective evolutionary algorithms (see Fig. 2.5) were analyzed, and some of them were related with the expected quality of the Pareto Front approximation:

- Convergence: Reaching the true and unknown Pareto Front.
- Diversity: Getting a useful spreading along the Pareto Front approximation.
- Pertinency: Obtaining useful and pertinent solutions for the designer.

While others were related with specific optimization instances, as:

- Constrained: usually non linear inequalities or equalities.
- Many-objectives: a problem with more than 3 design objectives to be optimized.
- Large scale: a problem with several decision variables (hundreds).
- Dynamic: where the main problem is tracking the Pareto Front, which is varying thought time.
- Expensive: a problem where its cost function calculation requires several CPU resources.
- Multimodal: a problem where several decision vectors point to the same objectives vector.
- Robust: based on looking for a suboptimal solution, if it guarantees its robustness.

G. Reynoso Meza et al., *Controller Tuning with Evolutionary Multiobjective Optimization*, Intelligent Systems, Control and Automation: Science and Engineering 85, DOI 10.1007/978-3-319-41301-3_3

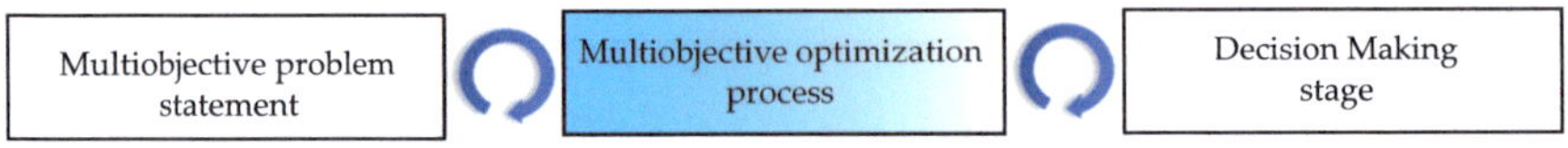

Fig. 3.1 The MO process in the MOOD procedure

Nowadays it is quite common to find algorithms that efficiently cover convergence and spreading properties. Nevertheless, in contrast, less work has been done regarding pertinency improvements. Pertinency could play a major role to close the gap between decision making and optimization, since it will be possible to get a more pertinent Pareto Front approximation for the designer. In this chapter, we will provide basic tools to handle such issues. The aim is to provide the reader with some basic tools to deal with the following sections and problem instances of this book. There are of course a wider range of tools that could be used for this purpose.

3.1.1 *Evolutionary Technique*

As basic evolutionary mechanism, the Differential Evolution (DE) algorithm is proposed [4, 16, 17]. Their usage is due to its simplicity and proved efficacy in several optimization instances. The most basic form will be used; it uses two operators: mutation and crossover (Eqs. (3.1) and (3.2) respectively) to generate its offspring (Algorithm 3.1, Fig. 3.2).

for *i=1:SolutionsInParentPopulation* **do**
    Generate a Mutant Vector $\boldsymbol{v}^i$ (Equation (3.1)) ;
    Generate a Child Vector $\boldsymbol{u}^i$ (Equation (3.2)) ;
end
Offspring $O = U$;

Algorithm 3.1: DE offspring generation mechanism.

Mutation: For each target (parent) vector $\boldsymbol{\theta}^i|_G$, a mutant vector $\boldsymbol{v}^i|_G$ is generated at generation G according to Eq. (3.1):

$$\boldsymbol{v}^i|_G = \boldsymbol{\theta}^{r_1}|_G + F(\boldsymbol{\theta}^{r_2}|_G - \boldsymbol{\theta}^{r_3}|_G). \tag{3.1}$$

where $r_1 \neq r_2 \neq r_3 \neq i$ are randomly selected, F is usually known as the scaling factor.

Crossover: For each target vector $\boldsymbol{\theta}^i|_G$ and its mutant vector $\boldsymbol{v}^i|_G$, a trial (child) vector $\boldsymbol{u}^i|_G = [u^i_1|_G, u^i_2|_G, \ldots, u^i_n|_G]$ is created as follows:

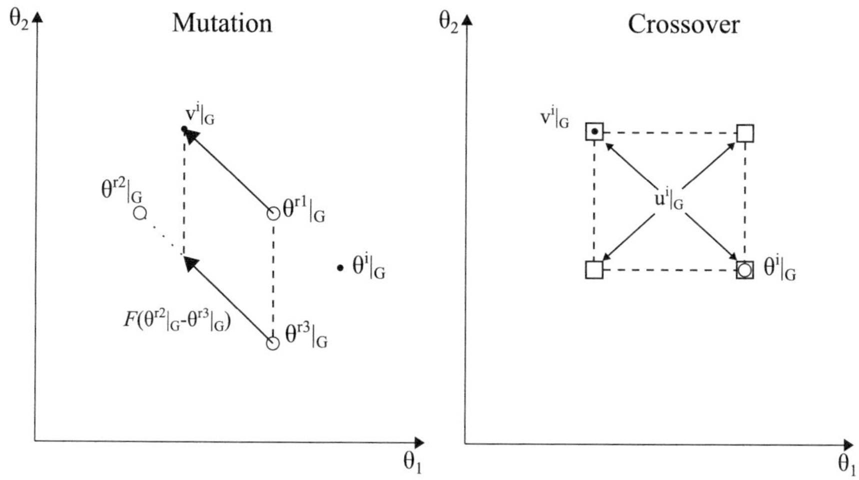

Fig. 3.2 DE operators (Mutation and Crossover) representation for a bi-dimensional searching space

$$u_j^i|_G = \begin{cases} v_j^i|_G & if\ rand(0,1) \leq Cr \\ \theta_j^i|_G & otherwise \end{cases} \tag{3.2}$$

where $j \in 1, 2, 3 \ldots n$ and Cr is named the crossover probability rate.

The standard selection mechanism is:

- A child is selected over its parent (for the next generation) if it has a better cost function value.

This mechanism selection is usually known as *greedy selection*. A pseudo-code for the basic DE is presented in Algorithm 3.2 and the tuning rules for its parameters are presented in Table 3.1.

Build initial population $P|_0$ with N_p individuals;
Evaluate $P|_0$;
Set generation counter $G = 0$;
while *stopping criterion unsatisfied* **do**
    $G = G + 1$;
    Build offspring $P^*|_G$ using $P|_{G-1}$ with DE algorithm operators (Algorithm 3.1);
    Evaluate offspring $P^*|_G$;
    Update population $P|_G$ with $P|_{G-1}$ and $P^*|_G$ using greedy selection mechanism;
end
Select the solution $\boldsymbol{\theta}^*$ from $P|_G$ with the best cost value;
RETURN $\boldsymbol{\theta}^*$;

Algorithm 3.2: DE algorithm pseudocode.

Table 3.1 Guidelines for DE's parameters tuning

Parameter	Value	Comments
F (scaling factor)	0.5	Recognized as good initial choice according to [17]
	[0.8, 1.0]	Values recognized for non-separable problems according to [4, 11]
Cr (crossover rate)	[0.1, 0.2]	Values recognized for separable problems according to [4, 11]
	0.5	Trade-off value for separable and non-separable problems. Default value used (for example) by MOEA/D algorithm [18]
N_p (population)	50	While a *five to ten times the number of decision variables* rule has been recognized as a thumb rule [17] for single objective optimization, here it is proposed a default size of 50 individuals

3.1.2 A MOEA with Convergence Capabilities: MODE

As commented in the last chapter, a common way to incorporate the simultaneous optimization approach in a single objective evolutionary algorithm, is using the dominance criteria as a selection mechanism. Following this idea, the selection operator in the basic DE is changed to:

- A child is selected over his parent if the child strictly dominates his parent (Definition 2.2).

With this idea, the same parameters of basic DE could be used (Table 3.1). In Algorithm 3.3 a pseudo-code of the MODE implementation is presented.

Build initial population $P|_0$ with N_p individuals;
Evaluate $P|_0$;
Set generation counter $G = 0$;
while *stopping criterion unsatisfied* **do**
  $G = G + 1$;
  Build offspring $P^*|_G$ using $P|_{G-1}$ with DE algorithm operators (Algorithm 3.1);
  Evaluate offspring $P^*|_G$;
  Update population $P|_G$ with $P^*|_G$ and $P|_{G-1}$ using greedy selection mechanism with dominance criteria (Definition 2.1);
end
Build Pareto set approximation $\boldsymbol{\Theta}_P^*$ by using $P|_G$;
RETURN Pareto set approximation $\boldsymbol{\Theta}_P^*$;

Algorithm 3.3: MODE algorithm pseudocode.

Obviously, this multiobjective differential evolution (MODE) algorithm does not guarantees spreading in the Pareto Front. It may approximate a single solution in

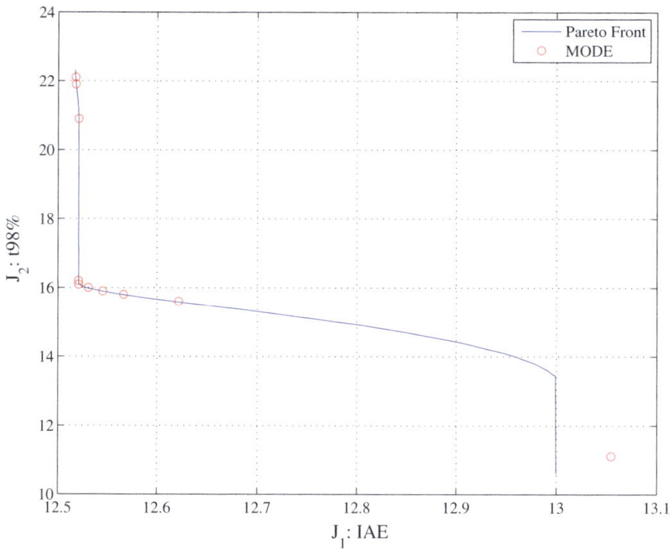

Fig. 3.3 Pareto front approximation with MODE algorithm for PI tunning example problem of Chap. 1

the Pareto Front (convergence) but it lacks any other proper mechanism to spread the solutions along the Pareto Front approximation. For our aforesaid example from Chap. 1, in Fig. 3.3 the Pareto Front approximation calculated for a single run of MODE algorithm is presented. You can notice that solutions converge quite well to the Pareto Front. Besides, it is observed that this approximation lacks proper spreading in order to cover all the Pareto Front. With this aim, a new mechanism to improve diversity will be added to this MODE algorithm.

3.1.3 An MODE with Diversity Features: sp-MODE

In order to promote diversity properties of the previous algorithm, a pruning mechanism will be added. A general pseudocode for MOEAs with pruning mechanism and external archive A is shown in Algorithm 3.4. The usage of an external archive A to store the best set of quality solutions found so far in an evolutionary process is a common practice in MOEAs.

With such purpose, in [13], a spherical pruning is proposed in order to attain a good distribution along the Pareto Front. The basic idea of the spherical pruning is to analyze the proposed solutions in the current Pareto Front approximation $\boldsymbol{J}_P^*$ by using normalized spherical coordinates from a reference solution (see Figs. 3.4 and 3.5). The pruning mechanism selects one solution for each spherical sector, according

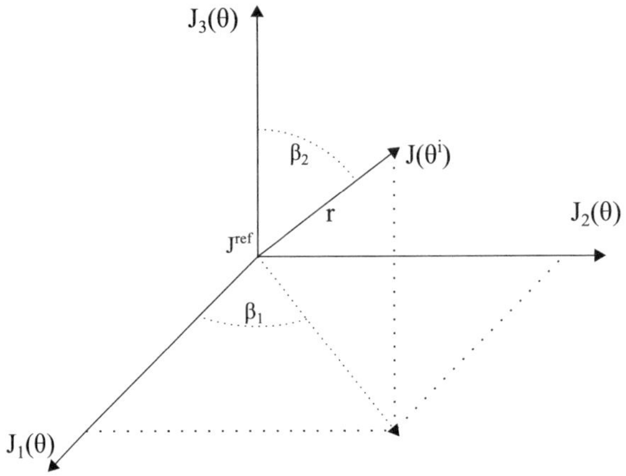

Fig. 3.4 Normalized spherical coordinate in a 3-dimensional space

Generate initial population $P|_0$ with N_p individuals;
Evaluate $P|_0$;
Apply dominance criterion (Definition 2.1) on $P|_0$ to get archive $\hat{A}|_0$;
Apply pruning mechanism to prune $\hat{A}|_0$ to get $A|_0$;
Set generation counter $G = 0$;
while *stopping criterion unsatisfied* **do**
  Update generation counter G=G+1;
  Get subpopulation $S|_G$ with solutions in $P|_{G-1}$ and $A|_{G-1}$;
  Generate offspring $O|_G$ with $S|_G$;
  Evaluate offspring $O|_G$;
  Update population $P|_G$ with offspring $O|_G$;
  Apply dominance criterion (Definition 2.1) on $O|_G \bigcup A|_{G-1}$ to get $\hat{A}|_G$;
  Apply pruning mechanism to prune $\hat{A}|_G$ to get $A|_G$;
  Update environment variables (if using a self-adaptive mechanism);
end
RETURN Pareto set approximation $\boldsymbol{\Theta}_P^* = A|_G$;

Algorithm 3.4: MOEA with pruning mechanism.

to a given norm or measure. This process is explained in Algorithm 3.5, where the following definitions are required:

Definition 3.1 (*Normalized spherical coordinates*) Given a solution $\boldsymbol{\theta}^i$ and $\boldsymbol{J}(\boldsymbol{\theta}^i)$, let

$$\boldsymbol{S}(\boldsymbol{J}(\boldsymbol{\theta}^i), \boldsymbol{J}^{ref}) = [r, \boldsymbol{\beta}] \tag{3.3}$$

be the normalized spherical coordinates from a reference point $\boldsymbol{J}^{ref}$ where $\boldsymbol{\beta} = [\beta_1, \ldots, \beta_{m-1}]$ is the arc vector and $r = \|\boldsymbol{J}(\boldsymbol{\theta}^i) - \boldsymbol{J}^{ref}\|_2$ is the Euclidean distance to the reference solution (see Figs. 3.4).

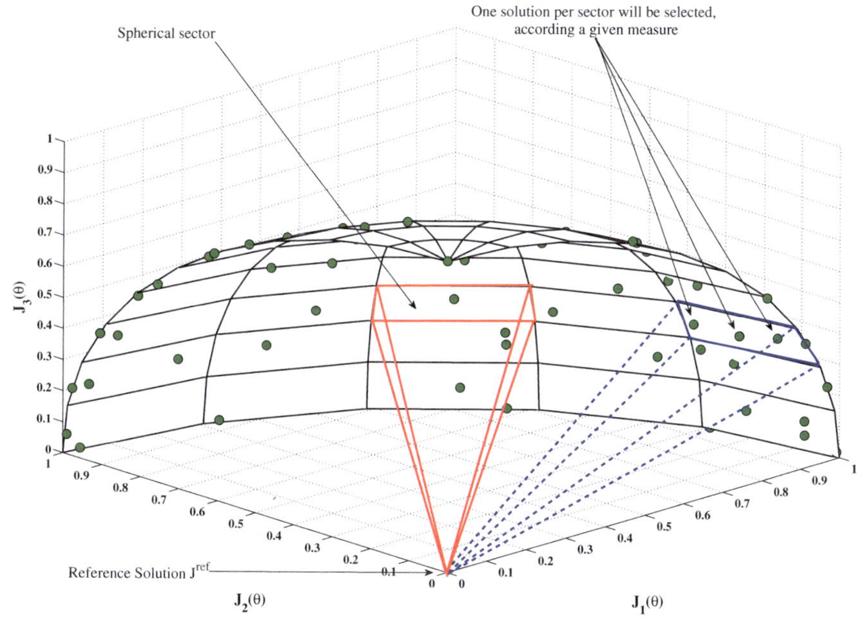

Fig. 3.5 Spherical relations on $\boldsymbol{J}_P^* \subset \mathbb{R}^3$. For each spherical sector, just one solution, the solution with the lowest norm will be selected

It is important to guarantee that $\boldsymbol{J}^{ref}$ dominates all the solutions. Given a Pareto Front approximation $\boldsymbol{J}_P^*$, an intuitive approach is to select

$$\boldsymbol{J}^{ref} = \boldsymbol{J}^{ideal} = \left[\min J_1(\boldsymbol{\theta}^i), \ldots, \min J_m(\boldsymbol{\theta}^i)\right] \forall \boldsymbol{J}(\boldsymbol{\theta}^i) \in \boldsymbol{J}_P^*. \tag{3.4}$$

Definition 3.2 (*Sight range*) The sight range from the reference solution $\boldsymbol{J}^{ref}$ to the Pareto Front approximation $\boldsymbol{J}_P^*$ is bounded by $\boldsymbol{\beta}^U$ and $\boldsymbol{\beta}^L$:

$$\boldsymbol{\beta}^U = \left[\max \beta_1(\boldsymbol{J}(\boldsymbol{\theta}^i)), \ldots, \max \beta_{m-1}(\boldsymbol{J}(\boldsymbol{\theta}^i))\right] \ \forall\ \boldsymbol{J}(\boldsymbol{\theta}^i) \in \boldsymbol{J}_P^*, \tag{3.5}$$

$$\boldsymbol{\beta}^L = \left[\min \beta_1(\boldsymbol{J}(\boldsymbol{\theta}^i)), \ldots, \min \beta_{m-1}(\boldsymbol{J}(\boldsymbol{\theta}^i))\right] \ \forall\ \boldsymbol{J}(\boldsymbol{\theta}^i) \in \boldsymbol{J}_P^*. \tag{3.6}$$

If $\boldsymbol{J}^{ref} = \boldsymbol{J}^{ideal}$, it is straightforward to prove that $\boldsymbol{\beta}^U = \left[\frac{\pi}{2}, \ldots, \frac{\pi}{2}\right]$ and $\boldsymbol{\beta}^L = [0, \ldots, 0]$.

Definition 3.3 (*Spherical grid*) Given a set of solutions in the objective space, the spherical grid on the m-dimensional space in arc increments $\boldsymbol{\beta}^\epsilon = [\beta_1^\epsilon, \ldots, \beta_{m-1}^\epsilon]$ is defined as:

$$\boldsymbol{\Lambda}^{\boldsymbol{J}_P^*} = \left[\frac{\beta_1^U - \beta_1^L}{\beta_1^\epsilon}, \ldots, \frac{\beta_{m-1}^U - \beta_{m-1}^L}{\beta_{m-1}^\epsilon}\right]. \tag{3.7}$$

Definition 3.4 (*Spherical sector*) The normalized spherical sector of a solution $\boldsymbol{\theta}^i$ is defined as

$$\boldsymbol{\Lambda}^{\epsilon}(\boldsymbol{\theta}^i) = \left[\left\lceil \frac{\beta_1(\boldsymbol{J}(\boldsymbol{\theta}^i))}{\Lambda_1^{J_P^*}} \right\rceil, \ldots, \left\lceil \frac{\beta_{m-1}(\boldsymbol{J}(\boldsymbol{\theta}^i))}{\Lambda_{m-1}^{J_P^*}} \right\rceil \right]. \tag{3.8}$$

Definition 3.5 (*Spherical pruning*) Given two solutions $\boldsymbol{\theta}^i$ and $\boldsymbol{\theta}^j$ from a set, $\boldsymbol{\theta}^i$ has preference in the spherical sector over $\boldsymbol{\theta}^j$ iff:

$$\left[\boldsymbol{\Lambda}^{\epsilon}(\boldsymbol{\theta}^i) = \boldsymbol{\Lambda}^{\epsilon}(\boldsymbol{\theta}^j)\right] \wedge \left[\|\boldsymbol{J}(\boldsymbol{\theta}^i) - \boldsymbol{J}^{ref}\|_p < \|\boldsymbol{J}(\boldsymbol{\theta}^j) - \boldsymbol{J}^{ref}\|_p\right] \tag{3.9}$$

where $\|\boldsymbol{J}(\boldsymbol{\theta}) - \boldsymbol{J}^{ref}\|_p = \left(\sum_{q=1}^{m} |J_q(\boldsymbol{\theta}) - J_q^{ref}|^p \right)^{1/p}$ is a suitable p-norm.

Read archive $\hat{A}|_G$;
Read and update extreme values for $\boldsymbol{J}^{ref}|_G$;
for *each member in* $\hat{A}|_G$ **do**
  calculate its normalized spherical coordinates (Definition 3.1);
end
Build the spherical grid (Definition 3.2 and 3.3);
for *each member in* $\hat{A}|_G$ **do**
  calculate its spherical sector (Definition 3.4);
end
for *i=1:SolutionsInArchive* **do**
  Compare with the remainder solutions in $\hat{A}|_G$;
  if *no other solution has the same spherical sector* **then**
    it goes to archive $A|_G$;
  end
  if *other solutions are in the same spherical sector* **then**
    it goes to archive $A|_G$ if it has the lowest norm (Definition 3.5);
  end
end
Return Archive $A|_G$;

Algorithm 3.5: Spherical pruning mechanism

The solution merging the MODE algorithm (Algorithm 3.3) and the spherical pruning mechanism (Algorithm 3.5) is known as sp-MODE[1] (see Algorithm 3.6). Default parameters and guidelines for parameter tuning are given in Table 3.2.

At this point, a MOEA with a diversity mechanism is available. In Fig. 3.6 a single run of sp-MODE (using the same number of function evaluations and parameters as in MODE algorithm) is presented for our PI tuning example. Notice that a better distribution is attained.

[1]Available at Matlab©Central (http://www.mathworks.com/matlabcentral/fileexchange/39215).

Generate initial population $P|_0$ with N_p individuals;
Evaluate $P|_0$;
Apply dominance criterion (Definition 2.1) on $P|_0$ to get $\hat{A}|_0$;
Apply pruning mechanism (Algorithm 3.5) to prune $\hat{A}|_0$ to get $A|_0$;
Set generation counter $G = 0$;
while *stopping criterion unsatisfied* **do**
    $G = G + 1$;
    Get subpopulation $S|_G$ with solutions in $P|_{G-1}$ and $A|_{G-1}$;
    Generate offspring $O|_G$ with $S|_G$ using DE operators (Algorithm 3.1).;
    Evaluate offspring $O|_G$;
    Update population $P|_G$ with offspring $O|_G$ according to greedy selection mechanism.;
    Apply dominance criterion (Definition 2.1) on $O|_G \bigcup A|_{G-1}$ to get $\hat{A}|_G$;
    Apply pruning mechanism (Algorithm 3.5) to prune $\hat{A}|_G$ to get $A|_G$;
end
RETURN Pareto set approximation $\boldsymbol{\Theta}_P^* = A|_G$;

Algorithm 3.6: sp-MODE algorithm pseudocode.

Table 3.2 Guidelines for sp-MODE's parameters tuning

Parameter	Value	Comments
DE algorithm		
F (scaling factor)	0.5	Recognized as good initial choice according to [17]
	[0.8, 1.0]	Values recognized for non-separable problems according to [4, 11]
Cr (crossover rate)	[0.1, 0.2]	Values recognized for separable problems according to [4, 11]
	0.5	Trade-off value for separable and non-separable problems. Default value used (for example) by MOEA/D algorithm [18]
N_p (population)	50	While a *five to ten times the number of decision variables* rule has been recognized as a thumb rule [17] for single objective optimization, here it is proposed a default size of 50 individuals

(continued)

Table 3.2 (continued)

Parameter	Value	Comments
Pruning mechanism		
	100	It is proposed for bi-objective problems, to bound the approximated Pareto front to 100 design alternatives
β^{ϵ} (Arcs)	[10, 10]	It is proposed for 3-objective problems, to bound the approximated Pareto front to $10^2 = 100$ design alternatives
	$[\overbrace{m, \ldots, m}^{m-1}]$	It is proposed for m-objective problems, to bound the approximated Pareto front to m^{m-1} design alternatives
p (p-norm)	1	It is proposed as default value

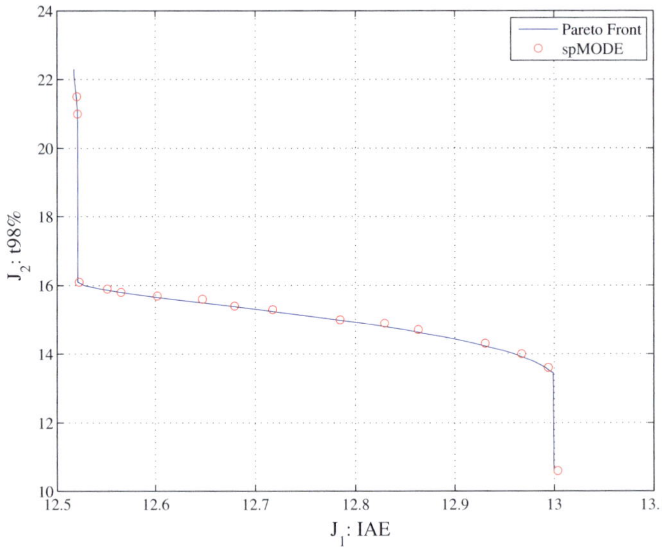

Fig. 3.6 Pareto Front approximation with sp-MODE algorithm (for PI tuning example problem of Chap. 1)

Even if you have an algorithm covering properly a Pareto Front, the designer may desire to focus on one certain region of the objectives exchange. That is, perhaps a solution is Pareto optimal, but it offers a strong degradation in one of the objectives, which makes this solution to be considered non-practical for the problem at hand. This idea will be explored in order to add an additional mechanism to improve the usability of the EMO process.

3.1.4 An sp-MODE with Pertinency Features: sp-MODE-II

A last final improvement over this algorithm concerns pertinency capabilities. For this purpose, a measure on the preferability of a solution will be incorporated in the pruning mechanism. By means of the Physical Programming (PP) method, such a preferability is calculated. PP is a suitable technique for multiobjective engineering design since it formulates design preferences in an understandable and intuitive language for designers. PP is an aggregate objective function (AOF) technique for multiobjective problems that includes available information in the optimization phase allowing the designer to express preferences for each objective with more detail. Firstly, PP translates the designers knowledge into classes with previously defined preference ranges. Preference sets reveal the DMs wishes using physical units for each of the objectives in the MOP. From this point of view, the problem is moved to a different space where all the variables are independent of the original MOP.

In [10] the PP methodology is modified, and a global PP (GPP) index is defined for a given objective vector. The main difference between PP and GPP is that the latter uses linear functions to built the class functions, while the former uses splines with several requirements to maintain convexity and continuity; the former fits better for local optimization algorithms, while the latter for (global) stochastic and evolutionary techniques. Thus, the GPP index will be used inside the sp-MODE's pruning mechanism.

Given an objective vector $\boldsymbol{J}(\boldsymbol{\theta}) = [J_1 \dots J_m]$, linear functions will be used for class functions $\eta_q(\boldsymbol{J})|_{\mathfrak{P}}$ as detailed in [14],[2] due to its simplicity and interpretability. Firstly, an offset between two adjacent ranges is incorporated (see Fig. 3.7) to meet the *one versus others* (OVO) rule criterion [1, 3].

Given M preference ranges for each one of the m objectives to manage, the preference matrix $\mathfrak{P}$ is defined as:

$$\mathfrak{P} = \begin{pmatrix} J_1^0 & \cdots & J_1^M \\ \vdots & \ddots & \vdots \\ J_m^0 & \cdots & J_m^M \end{pmatrix} \tag{3.10}$$

and the class functions $\eta_q(\boldsymbol{J})\big|_{\mathfrak{P}}$ are defined as:

$$\begin{aligned} \eta_q(\boldsymbol{J})\big|_{\mathfrak{P}} &= \alpha_{k-1} + \delta_{k-1} + \Delta\alpha_k \frac{J_q - J_q^{k-1}}{J_q^k - J_q^{k-1}} \\ J_q^{k-1} &\leq J_q < J_q^k \\ q &= [1, \dots, m] \\ k &= [1, \dots, M], \end{aligned} \tag{3.11}$$

[2] Hereafter, only 1S classes (the smaller, the better) will be considered.

where

$$\begin{aligned} \alpha_0 &= 0 \\ \Delta\alpha_k &> 0 \quad (1 < k \leq M) \\ \alpha_k &= \alpha_{k-1} + \Delta\alpha_k \quad (1 < k \leq M-1) \end{aligned} \tag{3.12}$$

$$\begin{aligned} \delta_0 &= 0 \\ \delta_k &> m \cdot (\alpha_k + \delta_{k-1}) - \alpha_k \quad (1 < k \leq M-1). \end{aligned} \tag{3.13}$$

The last inequality guarantees the *one versus others* (OVO) rule, since an objective value in a given range is always greater than the sum of the others in a most preferable range. Therefore, the GPP index, $J_{GPP}(\boldsymbol{J})$, is defined as:

$$J_{GPP}(\boldsymbol{J}) = \sum_{q=1}^{m} \eta_q(\boldsymbol{J})\big|_{\mathfrak{P}}. \tag{3.14}$$

The $J_{GPP}(\boldsymbol{J})$ index has an intrinsic structure to deal with constraints. If constraints fulfillment is required, they will be included in the preference set as objectives (preference ranges will be stated for each constraint and they will be used to compute the $J_{GPP}(\boldsymbol{J})$ index).

For example, class function $\eta_q(\boldsymbol{J})\big|_{\mathfrak{P}}$ representation is shown in Fig. 3.7 for the specific case (to be used hereafter) of five preference ranges defined as:

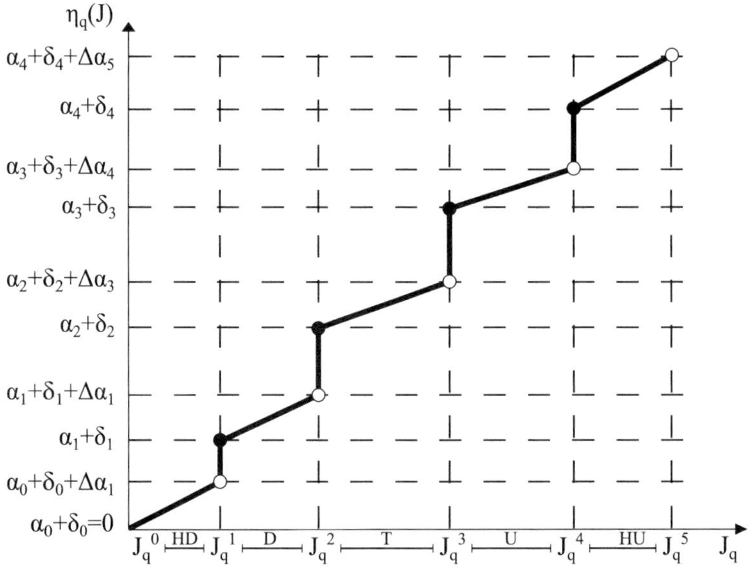

Fig. 3.7 Class definition for global physical programming

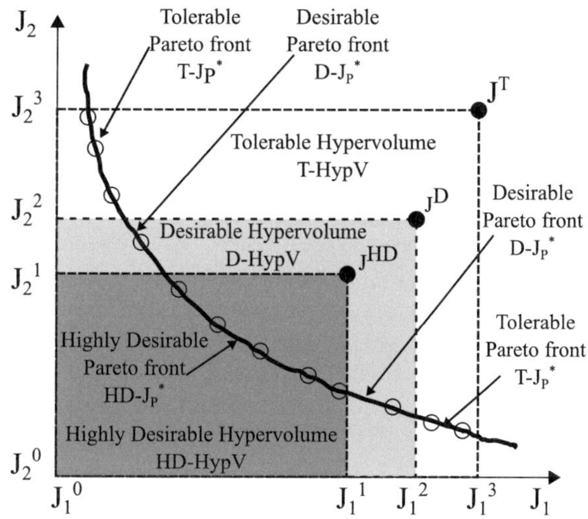

Fig. 3.8 Graphical representation of the definitions stated ($m = 2$)

HD: *Highly Desirable* if $J_q^0 \leq J_q < J_q^1$.
D: *Desirable* if $J_q^1 \leq J_q < J_q^2$.
T: *Tolerable* if $J_q^2 \leq J_q < J_q^3$.
U: *Undesirable* if $J_q^3 \leq J_q < J_q^4$.
HU: *Highly Undesirable* $J_q^4 \leq J_q < J_q^5$.

Those preference ranges are defined for the sake of flexibility (as it will be shown) to evolve the population to a pertinent Pareto Front. The following definitions will be used (see Fig. 3.8):

T_Vector: $\boldsymbol{J}^T = [J_1^3, J_2^3, \cdots, J_m^3]$, i.e., the vector with the maximum value for each objective in the *Tolerable* range.
D_Vector: $\boldsymbol{J}^D = [J_1^2, J_2^2, \cdots, J_m^2]$, i.e., the vector with the maximum value for each objective in the *Desirable* range.
HD_Vector: $\boldsymbol{J}^{HD} = [J_1^1, J_2^1, \cdots, J_m^1]$, i.e., the vector with the maximum value for each objective in the *Highly Desirable* range.
T_HypV: The hyper-volume of the Pareto front approximation bounded by $\boldsymbol{J}^T$.
D_Hypv: The hyper-volume of the Pareto Front approximation bounded by $\boldsymbol{J}^D$.
HD_HypV: The hyper-volume of the Pareto Front approximation bounded by $\boldsymbol{J}^{HD}$.
T_$\boldsymbol{J}_P^*$: The *Tolerable* Pareto Front approximation where all solutions dominate $\boldsymbol{J}^T$.
D_$\boldsymbol{J}_P^*$: The *Desirable* Pareto Front approximation where all solutions dominate $\boldsymbol{J}^D$.
HD_$\boldsymbol{J}_P^*$: The *Highly Desirable* Pareto Front approximation where all solutions dominate $\boldsymbol{J}^{HD}$.

Merging sp-MODE with Global Physical Programming (by means of Definition 3.14 and class functions with preference ranges of Fig. 3.7), the algorithm named sp-MODE-II is obtained. This algorithm keeps in each spherical sector the most preferable solution according to the DM's range of preferences. Furthermore, it could be used to prune and keep in a manageable size the Pareto Front approximation. That is, if the DM is looking to perform a MCDM stage with, for example, 100 solutions, you may request the algorithm to keep only the 100 best solutions (according to the GPP index) in the approximated Pareto Front. Algorithm 3.9 presents the pseudocode of the sp-MODE-II and default parameters and guidelines for their tuning are described in Table 3.3.

```
Read offspring (child population) O|_G and Subpopulation (parent) P|_{G-1};
for i=1:Solutions In Child Population do
    Get u^i from O|_G and θ^i from P|_{G-1};
    Calculate the physical index J_GPP(J(u^i)) and J_GPP(J(θ^i)) (Eq. 3.14);
    if J_GPP(J(θ^i)) > J_GPP^max then
        if J_GPP(J(u^i)) < J_GPP(J(θ^i)) then
            u^i goes to population P|_G
        else
            θ^i goes to population P|_G
        end
    else
        if u^i ⪯ θ^i then
            u^i goes to population P|_G
        else
            θ^i goes to population P|_G
        end
    end
end
Return Parent population P|_G;
```

Algorithm 3.7: DE selection procedure with global physical programming. J_{GPP}^{max} is an sp-MODE-II's parameter used to promote convergence towards the preferred area of DM (see details in Table 3.3).

In Fig. 3.9, a single run of the sp-MODE-II algorithm is presented for our PI tuning example, with preferences shown in Table 3.4.

Notice that the algorithm is achieving a spreading in the pertinent (*Tolerable*) region of the Pareto Front, avoiding uninteresting areas. Besides, this exchange in the number of solutions improves convergence properties, since the algorithm will focus in the interesting regions of the Pareto Front.

In any case, whether it has been used sp-MODE or sp-MODE-II, a decision-making process should be performed to select the preferred solution according to the stated preferences.

Read archive $\hat{A}|_G$ to be prune;
Read and update extreme values for $\boldsymbol{J}^{ref}|_G$;
for *each member in* $\hat{A}|_G$ **do**
    calculate its normalized spherical coordinates (Definition 3.1);
end
Build the spherical grid (Definition 3.2 and 3.3);
for *each member in* $\hat{A}|_G$ **do**
    calculate its spherical sector (Definition 3.4);
end
for *i=1:SolutionsInArchive* **do**
    if $J_{GPP}(\boldsymbol{J}(\boldsymbol{\theta}^i)) > J_{GPP}^{max}$ **then**
        $\boldsymbol{\theta}^i$ is not included in $A|_G$
    else
        Compare with the remainder solutions in $\hat{A}|_G$;
        if *no other solution has the same spherical sector* **then**
            it goes to the archive $A|_G$
        else
            it goes to the archive $A|_G$ if it has the lowest $J_{GPP}(\boldsymbol{J}(\boldsymbol{\theta}^i))$ (Eq. 3.14)
        end
    end
end
Return $A|_G$;

Algorithm 3.8: Spherical pruning with physical programming index.

Generate initial population $P|_0$ with N_p individuals;
Evaluate $P|_0$;
Apply dominance criterion (Definition 2.1) on $P|_0$ to get $\hat{A}|_0$;
Apply pruning mechanism based on $J_{GPP}(\boldsymbol{\theta})$ (Algorithm 3.8) to prune $\hat{A}|_0 G$ to get $A|_0$;
Set generation counter $G = 0$;
while *stopping criterion not satisfied* **do**
    $G = G + 1$;
    Get subpopulation $S|_G$ with solutions in $P|_{G-1}$ and $A|_{G-1}$;
    Generate offspring $O|_G$ with $S|_G$ using DE operators (Algorithm 3.1).;
    Evaluate offspring $O|_G$;
    Update population $P|_G$ with offspring $O|_G$ and $P|_{G-1}$ according to $J_{GPP}(\boldsymbol{J}(\boldsymbol{\theta}))$ values (Algorithm 3.7);
    Apply dominance criterion on $O|_G \bigcup A|_{G-1}$ to get $\hat{A}|_G$;
    Apply pruning mechanism based on $J_{GPP}(\boldsymbol{J}(\boldsymbol{\theta}))$ (Algorithm 3.8) to prune $\hat{A}|_G$ to get $A|_G$;
    Apply size control on $A|_{G+1}$, if apply.;
    Update environment variables (if using a self-adaptive mechanism);
end
RETURN Pareto set approximation $\boldsymbol{\Theta}_P^* = A|_G$;

Algorithm 3.9: sp-MODE-II algorithm pseudocode.

Table 3.3 Guidelines for sp-MODE-II's parameters tuning

Parameter	Value	Comments
DE algorithm		
F (scaling factor)	0.5	Recognized as good initial choice according to [17]
	[0.8, 1.0]	Values recognized for non-separable problems according to [4, 11]
Cr (crossover rate)	[0.1, 0.2]	Values recognized for separable problems according to [4, 11]
	0.5	Trade-off value for separable and non-separable problems. Default value used (for example) by MOEA/D algorithm [18]
N_p (population)	50	While a *five to ten times the number of decision variables* rule has been recognized as a thumb rule [17] for single objective optimization, here it is proposed a default size of 50 individuals
Pruning mechanism		
$\boldsymbol{\beta}^{\epsilon}$ (Arcs)	$10 \cdot \overbrace{[m, \ldots, m]}^{m-1}$	It is proposed for m-objective problems, to bound the grid size to m^{m-1} hyper spherical sectors
Pertinency mechanism		
J_{GPP}^{max}	$J_{GPP}(\boldsymbol{J}^T)$	It is proposed as default value. Only solutions with their m objectives in the Tolerable region could appear in $\boldsymbol{J}_P^*$
$\overline{car(J_P^*)}$	$10 \cdot m$	It is proposed as default value. In accordance to [7]
$\Delta\alpha_k$	$0.1, k > 0$	It is proposed in accordance to Eq. (3.12)
δ_k	$(m+1) \cdot (\alpha_k + \delta_{k-1}), k > 0$	It is proposed in accordance to Eq. (3.13)

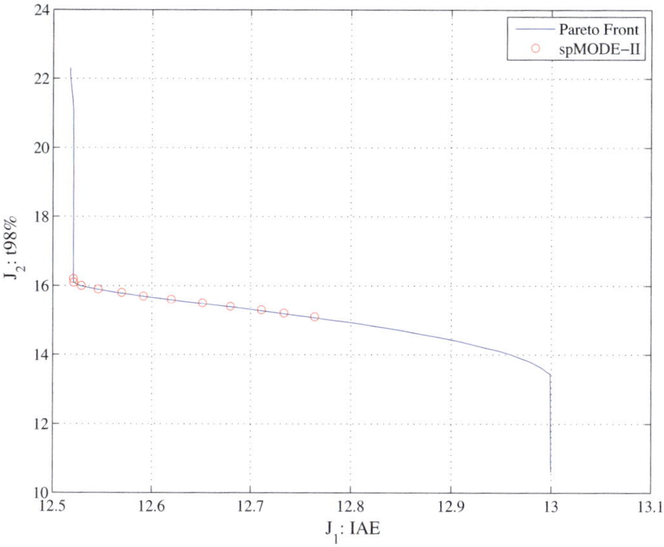

Fig. 3.9 Pareto front approximations with sp-MODE-II algorithm (for PI tunning example problem of Chap. 1)

Table 3.4 Preference matrix $\mathfrak{P}$ for PI tuning example. Five preference ranges have been defined ($M = 5$): *Highly Desirable* (HD), *Desirable* (D), *Tolerable* (T) *Undesirable* (U) and *Highly undesirable* (HU)

Objective	← HD →	← D →	← T →	← U →	← HU →	
	J_q^0	J_q^1	J_q^2	J_q^3	J_q^4	J_q^5
$J_1(\boldsymbol{\theta})$	12.0	12.5	12.7	12.8	12.9	13.0
$J_2(\boldsymbol{\theta})$	15.0	15.3	15.5	18.0	19.0	24.0

3.2 MCDM Stage

In this section, the third stage on MOOD procedure will be analyzed: the multicriteria decision making stage (Fig. 3.10). Starting from an approximation $\boldsymbol{J}_P^*$ to the real Pareto Front (obtained with sp-MODE, for example) composed of a large number of solutions in the m-dimensional space, the DM has to select the best solution according to his preferences. To make this decision, it is helpful to use mechanisms that include preferences easily and graphical tools for visualization and interactivity with data.

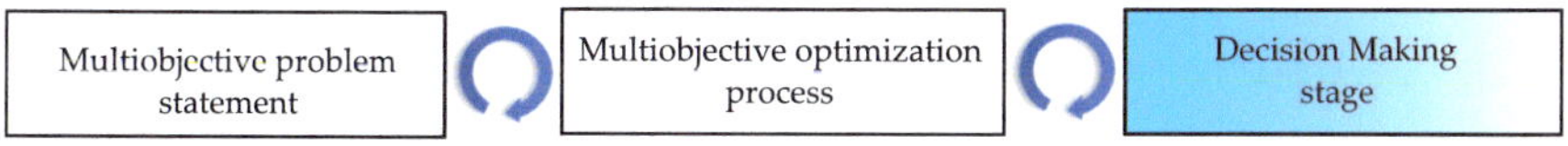

Fig. 3.10 The MCDM stage in the MOOD procedure

3.2.1 Preferences in MCDM Stage Using Utility Functions

The aim of the *Utility Functions* (sometimes called *Value Functions*) is to rank/classify Pareto points according to DM preferences. Ideally, preferences have to be expressed in a practical and meaningful way, directly connected with physical units of the problem. They depend on objectives, design parameters and any other information the DM could need to rank solutions.

There is a wide range of alternative to build *Utility Functions*, but in most cases a good selection is based on valuable characteristics as:

- An easy and intuitive way to transfer preferences.
- Meaningful preferences.

A simple example of *Utility Function* is to state the DM's preferences as a set of constraints over objectives. For instance, for a 3-D objective space ($\boldsymbol{J}(\boldsymbol{\theta}) = [J_1(\boldsymbol{\theta}), J_2(\boldsymbol{\theta}), J_3(\boldsymbol{\theta})]$), the DM prefers solutions that satisfy these four constraints:

$$\begin{aligned}
&\underline{J_1} \leq J_1(\boldsymbol{\theta}) \leq \overline{J_1}\\
&J_2(\boldsymbol{\theta}) \leq \overline{J_2}\\
&\underline{J_3} \leq J_3(\boldsymbol{\theta}).
\end{aligned}$$

If all constraints are equally important, an obvious and very simple *Utility Function* could be the number of constraints satisfied. If the constraints were: $1 \leq J_1(\boldsymbol{\theta}) \leq 10$, $J_2(\boldsymbol{\theta}) \leq 22$ and $5 \leq J_3(\boldsymbol{\theta})$, the preferred order for the following Pareto solutions would be (the higher utility value the better solution, it means more constraints satisfied):

$$\begin{aligned}
\boldsymbol{J}(\boldsymbol{\theta}^a) &= [7, 22, 6] \rightarrow \text{utility value: } 4\\
\boldsymbol{J}(\boldsymbol{\theta}^b) &= [2, 24, 7] \rightarrow \text{utility value: } 3\\
\boldsymbol{J}(\boldsymbol{\theta}^c) &= [-1, 25, 8] \rightarrow \text{utility value: } 2.
\end{aligned}$$

A more refined example can be the use of the aforementioned GPP index. Adjusting preferences for each objective consists of selecting the interval range for several labels that show the level of preference. Notice that these ranges are in physical units of the objectives (increasing meaningfulness). For an MOP of m objectives and setting $M = 3$ labels (*Desirable*, *Tolerable* and *Undesirable*), the DM must fulfill the Table 3.5.[3]

As the values of the table are in units of the design objectives they are understandable by the DM. The alternative to rank each Pareto solution could be focused on obtaining balanced solutions, that is, the solutions are ranked according to the worst label value of the (OVO-rule). So that $[T, T, \ldots, T, T]$ is preferred over $[D, D, \ldots, D, I]$.

[3]By means of this table, the DM is defining the matrix $\mathfrak{P}$ (Eq. 3.10).

Table 3.5 A general preference statement

	Desirable (D)	Tolerable (T)	Undesirable (U)
$J_1(\boldsymbol{\theta})$	$[J_1^0, J_1^1[$	$[J_1^1, J_1^2[$	$[J_1^2, J_1^3[$
$\vdots$	$\vdots$	$\vdots$	$\vdots$
$J_m(\boldsymbol{\theta})$	$[J_m^0, J_m^1[$	$[J_m^1, J_m^2[$	$[J_m^2, J_m^3[$

Table 3.6 Preferences for a greenhouse dynamic climate model

	Desirable (D)	Tolerable (T)	Undesirable (U)
$J_1(\boldsymbol{\theta}) = \|\|e_T\|\|_1$	$[0, 1[$°C	$[1, 3[$°C	$[3, 10[$°C
$J_2(\boldsymbol{\theta}) = \|\|e_T\|\|_\infty$	$[0, 5[$°C	$[5, 8[$°C	$[8, 20[$°C
$J_3(\boldsymbol{\theta}) = \|\|e_{HR}\|\|_1$	$[0, 5[$%	$[5, 15[$%	$[15, 30[$%
$J_4(\boldsymbol{\theta}) = \|\|e_{HR}\|\|_\infty$	$[0, 12[$%	$[12, 25[$%	$[25, 50[$%

For example, Table 3.6 defines the preferences for a MOP stated to identify a greenhouse dynamic climate model (temperature and relative humidity outputs) [5]. where $||e_T||_1$ and $||e_T||_\infty$ are the mean and maximum temperature identification errors[4] and $||e_{HR}||_1$ and $||e_{HR}||_\infty$ are the mean and maximum relative humidity errors respectively. Assuming the following Pareto solutions:

$$\begin{aligned}
\boldsymbol{\theta}^a &\rightarrow \boldsymbol{J}(\boldsymbol{\theta}^a) = [0.9, 4.5, 4.0, 28.0] \rightarrow [D, D, D, U] \\
\boldsymbol{\theta}^b &\rightarrow \boldsymbol{J}(\boldsymbol{\theta}^b) = [2.0, 5.5, 6.0, 15.0] \rightarrow [T, T, T, T] \\
\boldsymbol{\theta}^c &\rightarrow \boldsymbol{J}(\boldsymbol{\theta}^c) = [0.5, 12.5, 5.5, 14.0] \rightarrow [D, U, T, T] \\
\boldsymbol{\theta}^d &\rightarrow \boldsymbol{J}(\boldsymbol{\theta}^d) = [1.5, 7.5, 10.0, 24.5] \rightarrow [T, T, T, T]
\end{aligned}$$

they are preferred in the following order based on the stated preferences (from best to the worst):

$$\boldsymbol{\theta}^b, \boldsymbol{\theta}^d \rightarrow \boldsymbol{\theta}^a \rightarrow \boldsymbol{\theta}^c.$$

With the PP method and the OVO-rule you can not discriminate between solutions that belongs to the same rank in all objectives (as $\boldsymbol{\theta}^b$ and $\boldsymbol{\theta}^d$). Therefore a more accurate ranking is needed, as GPP. In Sect. 3.1.4, the sp-MODE-II was presented, including a GPP index to provide itself pertinence capabilities. Using it (Eq. 3.14) to rank solutions (Fig. 3.11):

$$J_{GPP}(\boldsymbol{J}) = \sum_{q=1}^{4} \eta_q(\boldsymbol{J})\big|_{\mathfrak{P}}$$

[4]error = simulated output - measured output.

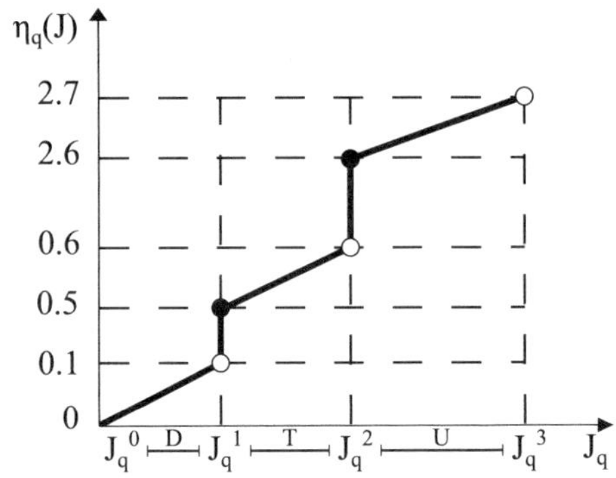

Fig. 3.11 Class definition in GPP for greenhouse model identification MOP

where:

$$\mathfrak{P} = \begin{pmatrix} J_1^0 = 0 & J_1^1 = 1 & J_1^2 = 3 & J_1^3 = 10 \\ J_2^0 = 0 & J_2^1 = 5 & J_2^2 = 8 & J_2^3 = 20 \\ J_3^0 = 0 & J_2^1 = 5 & J_3^2 = 15 & J_3^3 = 30 \\ J_4^0 = 0 & J_4^1 = 12 & J_4^2 = 25 & J_4^3 = 50 \end{pmatrix}$$

and

$$\eta_q(\boldsymbol{J})\big|_{\mathfrak{P}} = \alpha_{k-1} + \delta_{k-1} + \Delta\alpha_k \frac{J_q - J_q^{k-1}}{J_q^k - J_q^{k-1}}$$

$$J_q^{k-1} \leq J_q < J_q^k$$

$$q \in [1\dot{4}]$$

$$k \in [1 \ldots 3]$$

with[5]:

$$\alpha_0 = \delta_0 = 0$$

$$\Delta\alpha_1 = \Delta\alpha_2 = \Delta\alpha_3 = 0.1$$

$$\alpha_1 = 0.1; \alpha_2 = 0.2$$

$$\delta_1 = 0.5; \delta_2 = 2.8$$

Our greenhouse model identification MOP example gets (when $J_{GPP}(\boldsymbol{J})$ is applied to the previous $\boldsymbol{J}(\boldsymbol{\theta}^a)$, $\boldsymbol{J}(\boldsymbol{\theta}^b)$, $\boldsymbol{J}(\boldsymbol{\theta}^b)$ and $\boldsymbol{J}(\boldsymbol{\theta}^d)$):

$$J_{GPP}(\boldsymbol{J}(\boldsymbol{\theta}^a)) = 0.090 + 0.090 + 0.080 + 2.612 = 2.87$$

$$J_{GPP}(\boldsymbol{J}(\boldsymbol{\theta}^b)) = 0.550 + 0.517 + 0.510 + 0.523 = 2.10$$

$$J_{GPP}(\boldsymbol{J}(\boldsymbol{\theta}^c)) = 0.050 + 2.637 + 0.505 + 0.515 = 3.70$$

$$J_{GPP}(\boldsymbol{J}(\boldsymbol{\theta}^d)) = 0.525 + 0.583 + 0.550 + 0.596 = 2.25$$

[5]Recommendation from Table 3.3 has been followed to set $\Delta\alpha_k = 0.1$ and $\delta_k = m \cdot (\alpha_k + \delta_{k-1})$.

and the new rank of solutions, from best to worst, is:

$$\boldsymbol{\theta}^b \rightarrow \boldsymbol{\theta}^d \rightarrow \boldsymbol{\theta}^a \rightarrow \boldsymbol{\theta}^c.$$

This last discussion has been performed using just an *analytical approach* to rank an approximated Pareto Front. Nevertheless, merging such approaches with visualization tools may be useful for designers, in order to increment their embedment into the MCDM stage.

3.2.2 Level Diagrams for Pareto Front Analysis

Several visualization paradigms to depict m-dimensional data exists. Choosing one over another depends on the designer's needs. From a practical point of view, designers demand visualization technique that has three basic features:

- Analysing and comparing different design alternatives.
- Analysing and comparing different design concepts.
- Interacting with the depicted data.

According to the above characteristics, this book uses as a pivotal tool in the DM stage the visualization technique stated by Level Diagrams (LD). LD enables the DM to perform an analysis and classification on the approximated Pareto Front, $\boldsymbol{J}_P^*$, since each objective, $J_q(\boldsymbol{\theta})$, is normalized with respect to its minimum and maximum values. That is:

$$\hat{J}_q(\boldsymbol{\theta}) = \frac{J_q(\boldsymbol{\theta}) - J_q^{min}}{J_q^{max} - J_q^{min}}, q \in [1 \dots m] \tag{3.15}$$

where (with a little abuse of notation):

$$\boldsymbol{J}^{min} = \left[\min_{J_1(\boldsymbol{\theta}) \in J_P^*} J_1(\boldsymbol{\theta}), \dots, \min_{J_m(\boldsymbol{\theta}) \in J_P^*} J_m(\boldsymbol{\theta}) \right] \tag{3.16}$$

$$\boldsymbol{J}^{max} = \left[\max_{J_1(\boldsymbol{\theta}) \in J_P^*} J_1(\boldsymbol{\theta}), \dots, \max_{J_m(\boldsymbol{\theta}) \in J_P^*} J_m(\boldsymbol{\theta}) \right]. \tag{3.17}$$

For each normalized objective vector $\hat{\boldsymbol{J}}(\boldsymbol{\theta}) = [\hat{J}_1(\boldsymbol{\theta}), \dots, \hat{J}_m(\boldsymbol{\theta})]$, a p-norm $\|\hat{\boldsymbol{J}}(\boldsymbol{\theta})\|_p$ is applied to evaluate the distance to an ideal solution $\boldsymbol{J}^{ideal} = \boldsymbol{J}^{min}$. Common norms are:

$$\|\hat{\boldsymbol{J}}(\boldsymbol{\theta})\|_1 = \sum_{q=1}^{m} \hat{J}_q(\boldsymbol{\theta}) \tag{3.18}$$

$$\|\hat{\boldsymbol{J}}(\boldsymbol{\theta})\|_2 = \sum_{q=1}^{m} \hat{J}_q(\boldsymbol{\theta})^2 \tag{3.19}$$

$$\|\hat{\boldsymbol{J}}(\boldsymbol{\theta})\|_\infty = \max \hat{J}_q(\boldsymbol{\theta}). \tag{3.20}$$

The LD visualization deploys a two-dimensional graph for each objective and decision variable. The ordered pairs $\left(J_q(\boldsymbol{\theta}), \|\hat{\boldsymbol{J}}(\boldsymbol{\theta})\|_p\right)$ are plotted in each objective sub-graph and $\left(\theta_l, \|\hat{\boldsymbol{J}}(\boldsymbol{\theta})\|_p\right)$ in each decision variable sub-graph. Therefore, a given solution will have the same y-value in all graphs (see Fig. 3.12). This correspondence helps to evaluate general tendencies along the Pareto Front and to compare solutions according to the selected norm. In all cases, the lower the norm, the closer to the ideal solution.[6]

For example, an Euclidean norm $\|\cdot\|_2$ is helpful to evaluate the distance of a given solution with respect to the ideal solution, meanwhile a maximum norm will give information about the trade-off achieved by this solution. Such a norm, used to visualize tendencies in the Pareto Front, does not deform the MOP essence, since this visualization process takes place after the optimization stage.

In all cases, the lower the norm, the closer to the ideal solution $\boldsymbol{J}^{min}$. For example, in Fig. 3.12, point A is the closest solution to $\boldsymbol{J}^{min}$ with the $\|\cdot\|_2$ norm. It does not mean point A must be selected by the DM. Selection will be performed according to LD visualization and the DM's preferences. Also it is possible to visualize how the trade-off rate changes in solution A appreciating two different tendencies around that solution. On the one hand, the better $J_2(\boldsymbol{\theta})$ value, the worse $J_1(\boldsymbol{\theta})$ value. On the other hand, the worse $J_2(\boldsymbol{\theta})$ value, the better $J_1(\boldsymbol{\theta})$ one. Notice how it is difficult to appreciate such tendencies with classical visualizations with more than three dimensions.

The LD-ToolBox,[7] a powerful tool to analyze m-objective Pareto fronts, [8, 12, 19] is a Matlab © toolbox that offers to the DM a degree of interactivity with multidimensional data. An in-depth explanation of the LD tool capabilities can be found in [2].[8]

[6]In this book, the minimal values for each objective in the calculated Pareto Front approximation are used to build an ideal solution, $\boldsymbol{J}^{min}$.

[7]Available at http://www.mathworks.com/matlabcentral/fileexchange/24042.

[8]There are video tutorials available at http://cpoh.upv.es/es/investigacion/software/item/52-ld-tool.html.

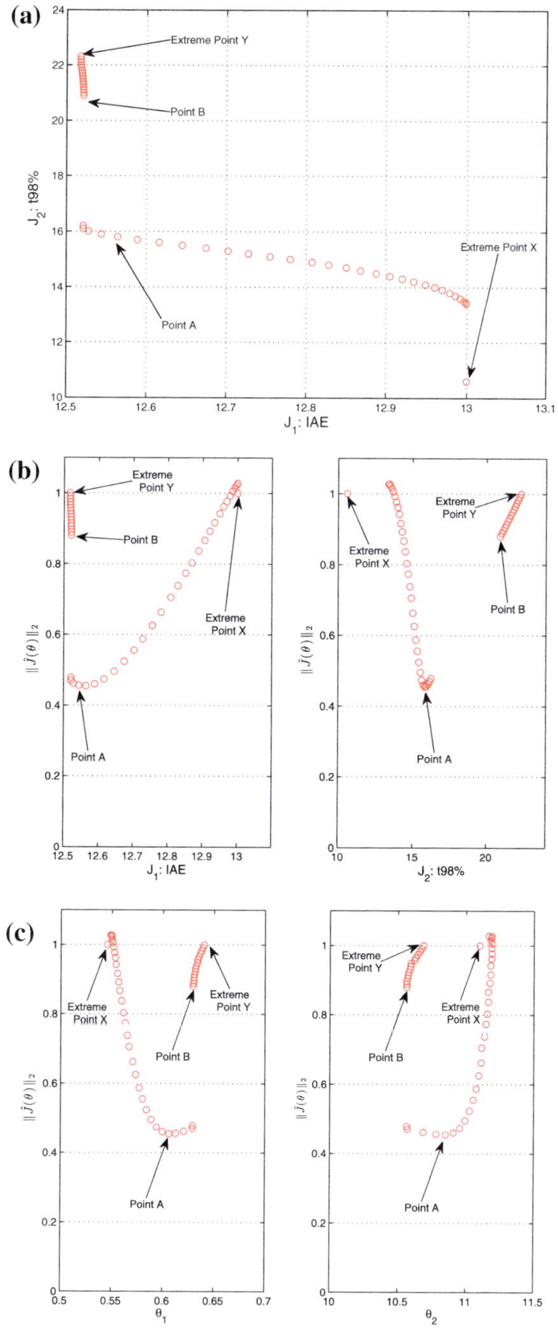

Fig. 3.12 Representation of the Pareto Front for our bi-objective problem using 2-D graph (**a**) and LD (**b**, **c**). Points at the same level in LD correspond on each graphic

3.2.3 Level Diagrams for Design Concepts Comparison

Further objectives trade-off analysis could include selection and comparison of various design concepts (i.e., different methods) for solving a MOP. The above examples were related with comparing trade-off of different design alternatives (solutions) for a given Pareto Front. Nevertheless, the designer could be interested in comparing trade-off surfaces of two or more Pareto Fronts (design concepts). For example, perhaps the designer is willing to compare the closed loop performance of a PID controller with the one achieved by a Fuzzy controller. An analysis of the objective exchange when different design concepts are used will provide a better insight into the problem at hand. This new analysis will help the DM to compare different design approaches, evaluating the circumstances where he/she would prefer one over another. Furthermore, the DM can decide whether the use of a complex concept is justified over a simple one. According to this, additional features for LD will be presented when design concepts comparison is needed. It is important to have in mind that:

- For the DM it is important to compare the degree of improvement of one design concept over other(s). This could be justified by the fact that some of the qualitative preferences of one design concept are important to bear in mind during the final selection. If there are no preferences for the design concepts under consideration, a Global Pareto Front could be calculated with solutions from all design concepts. In such case, the analysis on a single Pareto Front described in [2] with LD visualization would be enough.
- This visualization is complementary, i.e., it does not substitute the LD visualization technique shown in [2], but it gives additional information to the DM.

As pointed in [7], when multiple design concepts are evaluated by means of their Pareto Fronts, a measurement to quantify their weaknesses and strengths is needed. Both are essential to bring the usefulness of Pareto Fronts for conceptual design evaluation.

Several measurements have been developed to evaluate Pareto Front approximations. Nevertheless, many are incompatible or incomplete [20] with objective vector relations such as strict dominance, dominance or weak dominance (Definitions 2.1, 2.2, 2.3).

To evaluate relative performances between design concepts, the binary ϵ-indicator, I_ϵ, [6, 20] is used. This indicator shows the factor $I_\epsilon(\boldsymbol{J}^*_{Pi}, \boldsymbol{J}^*_{Pj})$ by which a set, $\boldsymbol{J}^*_{Pi}$, is worse than another, $\boldsymbol{J}^*_{Pj}$, with respect to all the objectives. As detailed in [20], this indicator is complete and compatible, and is useful to determine if two Pareto Fronts are incomparable, equal, or if one is better than the other (see Table 3.7 and Fig. 3.13).

Definition 3.6 The binary ϵ-indicator $I_\epsilon(\boldsymbol{J}^*_{P_i}, \boldsymbol{J}^*_{P_j})$ [20] for two Pareto front approximations $\boldsymbol{J}^*_{P_i}, \boldsymbol{J}^*_{P_j}$ is defined as:

Table 3.7 Interpretations for the I_ϵ indicator

$I_\epsilon(\boldsymbol{J}^*_{P_i}, \boldsymbol{J}^*_{P_j}) < 1$	$\rightarrow$	Every $\boldsymbol{J}(\boldsymbol{\theta}^j) \in \boldsymbol{J}^*_{P_j}$ is strictly dominated by at least one $\boldsymbol{J}(\boldsymbol{\theta}^i) \in \boldsymbol{J}^*_{P_i}$
$I_\epsilon(\boldsymbol{J}^*_{P_i}, \boldsymbol{J}^*_{P_j}) = 1 \wedge I_\epsilon(\boldsymbol{J}^*_{P_j}, \boldsymbol{J}^*_{P_i}) = 1$	$\rightarrow$	$\boldsymbol{J}^*_{P_i} = \boldsymbol{J}^*_{P_j}$
$I_\epsilon(\boldsymbol{J}^*_{P_i}, \boldsymbol{J}^*_{P_j}) > 1 \wedge I_\epsilon(\boldsymbol{J}^*_{P_j}, \boldsymbol{J}^*_{P_i}) > 1$	$\rightarrow$	Neither $\boldsymbol{J}^*_{P_i}$ weakly dominates $\boldsymbol{J}^*_{P_j}$ nor $\boldsymbol{J}^*_{P_j}$ weakly dominates $\boldsymbol{J}^*_{P_i}$

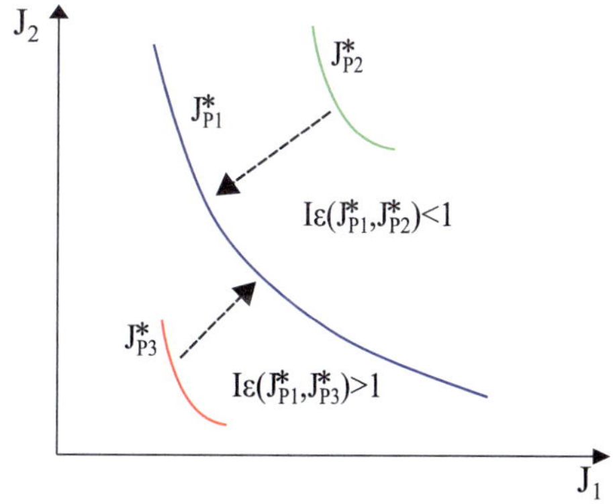

Fig. 3.13 Example of the binary ϵ-indicator to compare two pareto fronts

$$I_\epsilon(\boldsymbol{J}^*_{P_i}, \boldsymbol{J}^*_{P_j}) = \max_{\boldsymbol{J}(\boldsymbol{\theta}^j) \in \boldsymbol{J}^*_{P_j}} \epsilon(\boldsymbol{J}(\boldsymbol{\theta}^j), \boldsymbol{J}^*_{P_i}) \tag{3.21}$$

where

$$\epsilon(\boldsymbol{J}(\boldsymbol{\theta}^j), \boldsymbol{J}^*_{P_i}) = \min_{\boldsymbol{J}(\boldsymbol{\theta}^i) \in \boldsymbol{J}^*_{P_i}} \epsilon(\boldsymbol{J}(\boldsymbol{\theta}^i), \boldsymbol{J}(\boldsymbol{\theta}^j)) \tag{3.22}$$

$$\epsilon(\boldsymbol{J}(\boldsymbol{\theta}^i), \boldsymbol{J}(\boldsymbol{\theta}^j)) = \max_{1 \le l \le m} \frac{J_l(\boldsymbol{\theta}^i)}{J_l(\boldsymbol{\theta}^j)}, \tag{3.23}$$

As the binary ϵ-indicator is a scalar measure between two Pareto Fronts, some modifications are required to build a scalar measure for each design alternative on each design concept. The quality indicator Q[9] is defined for this purpose.

Definition 3.7 [9] The quality indicator $Q(\boldsymbol{J}(\boldsymbol{\theta}^i), \boldsymbol{J}^*_{P_j})$ for two design concepts $i, j \in [1, \ldots, K]$, $i \neq j$ and a design alternative $\boldsymbol{\theta}^i \in \boldsymbol{\Theta}^*_{P_i}, \boldsymbol{J}(\boldsymbol{\theta}^i) \in \boldsymbol{J}^*_{P_i}$ is defined as:

[9] To avoid problems with this quality indicator when the objective vector has positive, negative or zero values, a normalization in the range [1, 2] for each objective is used as a preliminary step.

Table 3.8 Comparison methods using the $Q(\boldsymbol{J}(\boldsymbol{\theta}^i), \boldsymbol{J}^*_{P_j})$ quality measure and its meaning

$Q(\boldsymbol{J}(\boldsymbol{\theta}^i), \boldsymbol{J}^*_{P_j}) < 1$	$\rightarrow$	$\boldsymbol{J}(\boldsymbol{\theta}^i) \in \boldsymbol{J}^*_{P_i}$ strictly dominates at least one $\boldsymbol{J}(\boldsymbol{\theta}^j) \in \boldsymbol{J}^*_{P_j}$	$\boldsymbol{J}(\boldsymbol{\theta}^i) \in \boldsymbol{J}^*_{P_i}$ has an improvement over a solution $\boldsymbol{J}(\boldsymbol{\theta}^j) \in \boldsymbol{J}^*_{P_j}$ by a scale factor of $Q(\boldsymbol{J}(\boldsymbol{\theta}^i), \boldsymbol{J}^*_{P_j})$ (at least) for all objectives
$Q(\boldsymbol{J}(\boldsymbol{\theta}^i), \boldsymbol{J}^*_{P_j}) = 1$	$\rightarrow$	$\boldsymbol{J}^i(\boldsymbol{\theta}^i) \in \boldsymbol{J}^*_{Pi}$ is not comparable with any solution $\boldsymbol{J}(\boldsymbol{\theta}^j) \in \boldsymbol{J}^*_{P_j}$	$\boldsymbol{J}(\boldsymbol{\theta}^i) \in \boldsymbol{J}^*_{P_i}$ is pareto optimal in $\boldsymbol{J}^*_{P_j}$ or $\boldsymbol{J}(\boldsymbol{\theta}^i) \in \boldsymbol{J}^*_{P_i}$ is inside a region in the objective space not covered by $\boldsymbol{J}^*_{P_j}$
$Q(\boldsymbol{J}(\boldsymbol{\theta}^i), \boldsymbol{J}^*_{P_j}) > 1$	$\rightarrow$	$\boldsymbol{J}(\boldsymbol{\theta}^i) \in \boldsymbol{J}^*_{P_i}$ is strictly dominated by at least one $\boldsymbol{J}(\boldsymbol{\theta}^j) \in \boldsymbol{J}^*_{P_j}$	A solution $\boldsymbol{J}(\boldsymbol{\theta}^j) \in \boldsymbol{J}^*_{Pj}$ has an improvement over $\boldsymbol{J}(\boldsymbol{\theta}^i) \in \boldsymbol{J}^*_{P_i}$ by a scale of $Q(\boldsymbol{J}(\boldsymbol{\theta}^i), \boldsymbol{J}^*_{P_j})$ (at least) for all objectives

$$Q(\boldsymbol{J}(\boldsymbol{\theta}^i), \boldsymbol{J}^*_{P_j}) = \begin{cases} 1 & \text{if } \epsilon(\boldsymbol{J}(\boldsymbol{\theta}^i), \boldsymbol{J}^*_{P_j}) > 1 \\ & \wedge \\ & \epsilon(\boldsymbol{J}(\boldsymbol{\theta}^j), \boldsymbol{J}^*_{P_i}) > 1 \ \forall \boldsymbol{J}(\boldsymbol{\theta}^j) \in \boldsymbol{J}^*_{P_j}. \\ \epsilon(\boldsymbol{J}(\boldsymbol{\theta}^i), \boldsymbol{J}^*_{P_j}) & \text{otherwise} \end{cases} \tag{3.24}$$

Combining LD visualization with the quality indicator, regions in the Pareto Front where a design concept is better or worse than another can be localized, offering a measurement of how much better one design concept performs than the other (see Table 3.8).

Lets assume we would like to compare the set of controllers $\boldsymbol{\Theta}_{P1}$ of our previous example[10] (design concept 1) with the SIMC PID tuning rules [15] for a FOPDT process (design concept 2):

$$K_c = \frac{T + L/3}{K(\tau_{cl} + L)} \tag{3.25}$$

$$T_i = \min\,(T + L/3, 4(\tau_{cl} + L)) \tag{3.26}$$

where $T = 10$ is the time constant, $L = 3$ the system delay, $K = 3.2$ the process gain and τ_{cl} the desired closed-loop time constant. By varying the parameter τ_{cl} it is possible to calculate a set of controllers $\boldsymbol{\Theta}_{P2}$ with different performance and robustness trade-off. We will compare both sets of controllers $\boldsymbol{\Theta}_{P1}$ and $\boldsymbol{\Theta}_{P2}$ with the design objectives $J_{ISE}(\boldsymbol{\theta})$ and $J_{IAU}(\boldsymbol{\theta})$ (Eqs. (2.11) and (2.17) respectively) for

[10] $\boldsymbol{\Theta}_1$ includes the Pareto Set of Table 1.4 obtained for IAE and $t_{98\,\%}$ minimization.

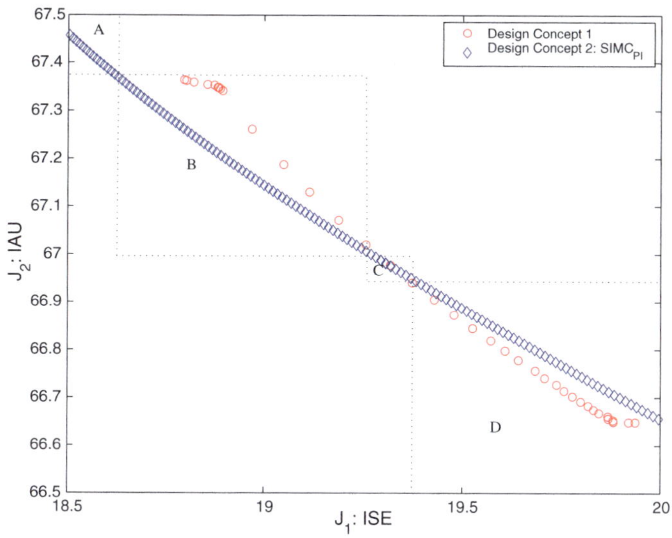

Fig. 3.14 Typical comparison of two design concepts using a 2-D graph. *A*, *B*, *C* and *D* identified areas by means of quality indicator Q (see Fig. 3.15b)

a setpoint step change. After computing both objectives and filtering dominated solutions, the Pareto set approximations $\boldsymbol{\Theta}^*_{P1}, \boldsymbol{\Theta}^*_{P2}$ and their respective Pareto Fronts $\boldsymbol{J}^*_{P1}, \boldsymbol{J}^*_{P2}$ are obtained (Fig. 3.14).

In Fig. 3.15 both Pareto Fronts (design concepts) in LD are depicted where the relationships described in Table 3.8 can be seen. Firstly, due to the quality indicator, it is possible to quickly identify the s-Pareto non-optimal (any solution $Q(\boldsymbol{J}(\boldsymbol{\theta}^i), \boldsymbol{J}^*_{P_j}) > 1$) from s-Pareto optimal (Definition 2.7) solutions (any solution $Q(\boldsymbol{J}(\boldsymbol{\theta}^i), \boldsymbol{J}^*_{P_j}) \leq 1$). Moreover, the quality indicator assigns a quantitative value about how better or worse a solution is with respect another concept. Further analysis with the quality indicator can be made for particular solutions or for regions in the LD.

Two particular solutions (design alternatives), $\boldsymbol{J}(\boldsymbol{\theta}^b) \in \boldsymbol{J}^*_{P1}$, and $\boldsymbol{J}(\boldsymbol{\theta}^a) \in \boldsymbol{J}^*_{P2}$, have been remarked in Fig. 3.15b. Notice that:

- $Q(\boldsymbol{J}(\boldsymbol{\theta}^a), \boldsymbol{J}^*_{P1}) \approx 0.95$. That is, among the solutions $\boldsymbol{J}(\boldsymbol{\theta}^1) \in \boldsymbol{J}^*_{P1}$ dominated by objective vector $\boldsymbol{J}(\boldsymbol{\theta}^a)$, the bigger k for a solution $\boldsymbol{J}(\boldsymbol{\theta}^1)$ such that $\boldsymbol{J}'(\boldsymbol{\theta}^1) = k \cdot \boldsymbol{J}(\boldsymbol{\theta}^1)$ weakly dominates $\boldsymbol{J}(\boldsymbol{\theta}^a)$ is $k \approx 0.95$.
- $Q(\boldsymbol{J}(\boldsymbol{\theta}^b), \boldsymbol{J}^*_{P2}) \approx 1.04$. That is, among the solutions $\boldsymbol{J}(\boldsymbol{\theta}^2) \in \boldsymbol{J}^*_{P2}$ which dominate objective vector $\boldsymbol{J}(\boldsymbol{\theta}^b)$, the smaller k for a solution $\boldsymbol{J}(\boldsymbol{\theta}^2)$ such that $\boldsymbol{J}'(\boldsymbol{\theta}^2) = k \cdot \boldsymbol{J}(\boldsymbol{\theta}^2)$ is weakly dominated by $\boldsymbol{J}(\boldsymbol{\theta}^b)$ is $k \approx 1.04$.

Regarding zones in Fig. 3.15b, zone B represents where design concept 2 (◇) is better than design concept 1 (○). Notice that, for zone B, the design alternatives from concept 2 have a quality measurement $Q(\boldsymbol{J}(\boldsymbol{\theta}^2), \boldsymbol{J}^*_{P1}) < 1$ and design alternatives from concept 1 have a quality measurement $Q(\boldsymbol{J}(\boldsymbol{\theta}^1), \boldsymbol{J}^*_{P2}) > 1$. The opposite is true

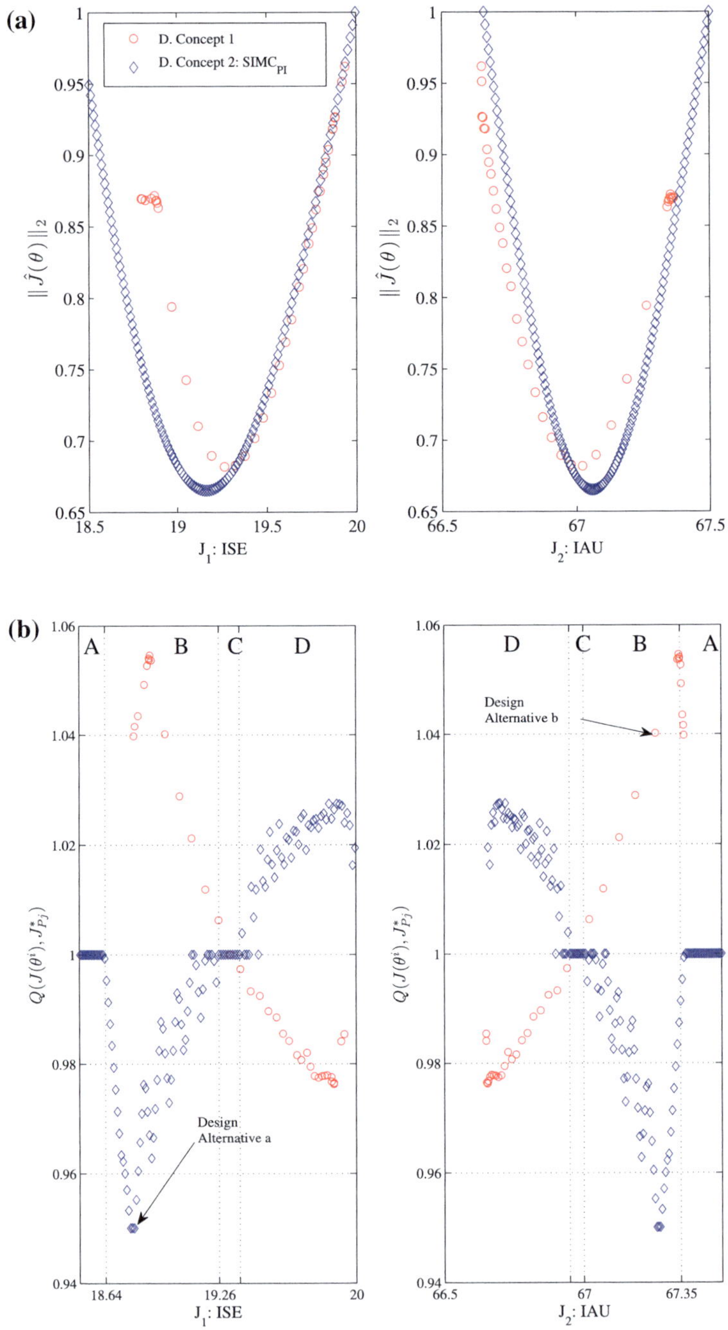

Fig. 3.15 Comparison of two design concepts: **a** typical level diagrams with 2-norm, **b** level diagrams with quality indicator Q

for zone D. Zone A is reached (covered) only by concept 2 (and thus, it is impossible to compare both concepts). Finally, in zone C both concepts have almost the same exchange between objectives. Concluding this, would be more difficult just analyzing an LD with standard norms (see Fig. 3.15a).

Although it is possible to build an s-Pareto Front merging the design alternatives of each concept and to analyze its tendencies, it would be very difficult to measure the improvement of one concept over another. This is mainly due to the loss of information after building the s-Pareto Front. Therefore the LD with the quality indicator enables a quantitative a-priori analysis between concepts, and it makes possible to decide, for example, if the improvement of one of them is significant or not. While such comparison can be performed by visual inspection in a classical 2D-objective graph (see Fig. 3.14), such a task will be more complex when three or more objectives are considered.

The design concepts comparison allows us also to reinforce the idea and philosophy behind the MOOD procedure for controller tuning applications. On the one hand, if a set is Pareto-optimal (Design concept 1) for a given pair of design objectives, that does not imply that it will be Pareto-optimal when the design objectives change; here relies the importance of stating (correctly) meaningful design objectives for the designer. On the other hand, two or three design objectives could not be enough to represent properly the expected behavior of a controller. So, here we emphasize again the main hypothesis about when this (book) procedure will be valuable for the designer:

- We use the MOOD procedure because it is difficult to find a controller with a reasonable balance among design objectives.
- We use the MOOD procedure because it is worthwhile analyzing the trade-off among controllers (design alternatives or design concepts).

3.3 Conclusions of This Chapter

This chapter is dedicated to presenting the tools that will be used throughout the book to solve MOOD procedures. Related to the optimization process, the sp-MODE algorithm will be used to obtain an approximation to the Pareto Front of a MO problem. Thanks to the properties of sp-MODE, approximations with good convergence and diversity will be achieved, so that the DM will have enough alternatives to choose the desired final solution.

sp-MODE-II algorithm will be used including design preferences exploiting its pertinence property. Then the algorithm will focus all its efforts on the area of interest, getting more interesting solutions to the DM.

Finally, in the MCDM stage, for the m-dimensional case ($m > 2$), LD graphical tool will be used, taking profit of its flexibility and graphics performance to choose the solution to the MOP.

References

1. Blasco, X., García-Nieto S, Reynoso-Meza G (2012) Autonomous trajectory control of a quadricopter vehicle. Simulation and evaluation. Revista Iberoamericana de Automática e Informática Industrial 9(2):194–199
2. Blasco X, Herrero J, Sanchis J, Martínez M (2008) A new graphical visualization of n-dimensional Pareto front for decision-making in multiobjective optimization. Inf Sci 178(20):3908–3924
3. Blasco X, Reynoso-Meza G, García-Nieto S (2013) Resultados del concurso de ingeniería de control 2012 y convocatoria 2013. Revista Iberoamericana de Automática e Informática Industrial 10(2):240–244
4. Das S, Suganthan PN (2010) Differential evolution: a survey of the state-of-the-art. IEEE Trans Evol Comput 99:1–28
5. Herrero J, Blasco X, Martínez M, Ramos C, Sanchis J (2008) Robust identification of non-linear greenhouse model using evolutionary algorithms. Control Eng Pract 16:515–530
6. Knowles J, Thiele L, Zitzler E (2006) A tutorial on the performance assessment of stochastic multiobjective optimizers. Tech. Rep. TIK report No. 214, Computer Engineering and networks laboratory. ETH Zurich, 2006
7. Mattson CA, Messac A (2005) Pareto frontier based concept selection under uncertainty, with visualization. Optim Eng 6:85–115
8. Reynoso-Meza G, Blasco X, Sanchis J (2009) Multi-objective design of PID controllers for the control benchmark 2008–2009 (in spanish). Revista Iberoamericana de Automática e Informática Industrial 6(4):93–103
9. Reynoso-Meza G, Blasco X, Sanchis J, Herrero JM (2013) Comparison of design concepts in multi-criteria decision-making using level diagrams. Inf Sci 221:124–141
10. Reynoso-Meza G, Sanchis J, Blasco X, García-Nieto S (2014) Physical programming for preference driven evolutionary multi-objective optimization. Appl Soft Comput 24:341–362
11. Reynoso-Meza G, Sanchis J, Blasco X, Herrero J (2011) Hybrid DE algorithm with adaptive crossover operator for solving real-world numerical optimization problems. In: 2011 IEEE congress on evolutionary computation (CEC) (June 2011), pp 1551–1556
12. Reynoso-Meza G, Sanchis J, Blasco X, Herrero JM (2012) Multiobjective evolutionary algortihms for multivariable PI controller tuning. Expert Syst Appl 39:7895–7907
13. Reynoso-Meza G, Sanchis J, Blasco X, Martínez M (2010) Design of continuous controllers using a multiobjective differential evolution algorithm with spherical pruning. In: Applications of evolutionary computation. Springer, pp 532–541
14. Sanchis J, Martínez MA, Blasco X, Reynoso-Meza G (2010) Modelling preferences in multi-objective engineering design. Eng Appl Artif Intell 23:1255–1264
15. Skogestad S, Grimholt C (2012) The simc method for smooth pid controller tuning. In: PID control in the third millennium. Springer, pp 147–175
16. Storn R (2008) SCI: Differential evolution research: Trends and open questions. vol LNCS 143. Springer, Heidelberg, pp 1–31
17. Storn R, Price K (1997) Differential evolution: a simple and efficient heuristic for global optimization over continuous spaces. J Global Optim 11:341–359
18. Zhang Q, Li H (2007) MOEA/D: a multiobjective evolutionary algorithm based on decomposition. IEEE Trans Evol Comput 11(6):712–731
19. Zio E, Razzo R (2010) Multiobjective optimization of the inspection intervals of a nuclear safety system: a clustering-based framework for reducing the pareto front. Ann Nuclear Energy 37:798–812
20. Zitzler E, Thiele L, Laumanns M, Fonseca C, da Fonseca V (2003) Performance assessment of multiobjective optimizers: an analysis and review. IEEE Trans Evol Comput 7(2):117–132

Part II
Basics

This part is dedicated to present basic examples regarding the multiobjective optimization design (MOOD) procedure for controller tuning. With such examples, basic and general optimization statements for univariable and multivariable processes are stated. The aim of such examples is to provide to practitioners a starting point to use the MOOD procedure in their own optimization instances.

Chapter 4
Controller Tuning for Univariable Processes

Abstract In this chapter, a simple controller tuning statement by means of the multiobjective optimization design procedure is given. The aim of this chapter is to show a basic example in controller tuning and to focus on the importance of the chosen objectives and the basic use of Level Diagrams as tool for the decision making process. Additionally the multiobjective approach will be compared with available tuning rules, in order to evaluate the tradeoff among those solutions, showing how multiobjective optimization tools could be used.

4.1 Introduction

In general, different factors should be considered when tuning a controller, but depending on the type of application some of them could be more important than others. Most common factors are related with:

- Set-point response (dynamic and steady-state closed loop behavior).
- Load disturbances.
- Process uncertainties.
- Noise.

It is easy to find in the literature different ways to quantify the attained performances (most common have been already presented in Chap. 2). In this chapter the designer wants a better understanding of the trade-off between the different objectives, therefore MOOD is worthwhile. Although the selection of the objectives is one key element for a satisfactory problem resolution, some conventional performance indicators will be used to allow a comparison with other tuning methodologies present in the literature.

G. Reynoso Meza et al., *Controller Tuning with Evolutionary Multiobjective Optimization*, Intelligent Systems, Control and Automation: Science and Engineering 85, DOI 10.1007/978-3-319-41301-3_4

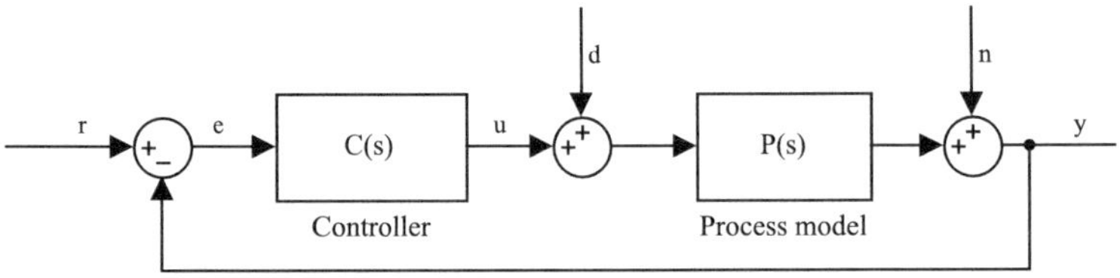

Fig. 4.1 Basic loop for PI control

4.2 Model Description

A high order process with delay is selected, represented by transfer function (4.1). For this case (complexity in the model), achievable performance depends highly on the degrees of freedom (DOF) of the controller $C(s)$, but a 1-DOF PI controller over error signal will be used (Fig. 4.1).

$$P(s) = \frac{1}{(s+1)^3} e^{-3s}, \tag{4.1}$$

$$C(s) = K_c \left(1 + \frac{1}{T_i \cdot s}\right). \tag{4.2}$$

4.3 The MOOD Approach

Indicators selected for set-point response performance evaluation are settling time at 98 % of steady state value ($J_{t98\,\%}$) and percentage of overshoot (J_{over}). These type of indicators are interpreted easier by the designer than other classical indicators as IAE or ITAE since minimizing these objectives means a fast closed loop response with low overshoot.

If rejection of load disturbances are required, an additional objective is added to the design procedure. An intuitive selection is minimizing the maximum deviation (in units of the controlled variable) produced by a unitary step change in the load disturbance (J_{overd}) (d in Fig. 4.1).

Focusing only in these three indicators is a three dimensional MOP with only two decision variables K_c and T_i (parameters of the PI controller). If constraints on these parameters $\boldsymbol{\theta} = [K_c, T_i]$ are set in the ranges: $K_c = [0.1, 2]$ and $T_i = [0.1, 6]$ and an additional stability constraint is added as a penalty function in the objectives functions in order to avoid unstable solutions, the MOP can be stated as:

$$\min_{\boldsymbol{\theta}}[J_{t98\,\%}(\boldsymbol{\theta}), J_{over}(\boldsymbol{\theta}), J_{overd}(\boldsymbol{\theta})] \tag{4.3}$$

$$st:$$

$$0.1 \leq K_c \leq 2$$
$$0.1 \leq T_i \leq 6$$
$$\text{Stable in closed loop.}$$

where closed loop stability is calculated from frequency margins using *allmargins* function of Matlab©Control Toolbox) and settling time and overshoot are calculated by simulation of the control loop using a Simulink©file built for this purpose.

The optimization algorithm used is sp-MODE-II adjusted as follows: spherical pruning with Euclidean norm and 50 arcs, front size limited to 100 solutions, function evaluations limited to 50,000. With this configuration, the algorithm has found a Pareto front approximation with 86 solutions.

Traditionally, the $\|\boldsymbol{J}\|_2$ norm is used for y-axis LD synchronization presenting a geometrical-like visualization, but for trade-off analysis, it is better to use a norm where y-axis interpretation supplies some additional information for decision making. Therefore, if $\infty-$norm ($\|\boldsymbol{J}\|_\infty$), the limit of y-axis in every figure of the LD is always 1. The points with $\|\boldsymbol{J}\|_\infty = 1$ have at least one of its components (objectives) in an extreme value of the Pareto Front. Interpretation for the rest of values is quite understandable, it shows the normalized distance (between 0 and 1) of the worst objective value for a particular point. For instance, a solution with $\|\boldsymbol{J}\|_\infty = 0.5$ means that the worst component of its objective vector is at the 50 % of the span for this objective. LD will show the value of the worst objective for this particular solution at the middle of the scale (x-axis).

Figure 4.2 shows the Pareto Front and Set obtained, using LD with $\|\boldsymbol{J}\|_\infty$ for y-axis synchronization. For better interpretation, each solution is colored in each LD figure in a same color. A quick view of this figure shows some features of the Pareto Front/Set obtained:

- $J_{t98\,\%}$ varies from values under 20 s to the limit of the simulation (100 s) (see x-axis limits at the $J_{t98\,\%}$ Level Diagram (Fig. 4.2)).
- J_{over} varies from 0 to 0.8. Clearly there are lots of solutions with very high overshoot, for instance from 0.2 to 0.8 (20–80 % of overshoot), but solutions with lower overshoot (for instance under 20 %) get good settling times (under 50 s) (see zone A in $J_{t98\,\%}$ and J_{over} diagrams).
- J_{overd} is clearly in conflict with the other two objectives $J_{t98\,\%}$ and J_{over}. Admissible values for $J_{t98\,\%}$ and J_{over} produce the poorest values of J_{overd} (see zone A at J_{overd} diagram), but the range of values for this objective is quite narrow, from around 0.86 to near 0.90 (x-axis of J_{overd} diagram). A deeper analysis of this LD shows that low improvements of J_{overd} (form 0.87 to 0.86) produce far worse $J_{t98\,\%}$ and J_{over} values (see zone B at the diagrams).
- From the Pareto set, admissible solutions (located previously at zone A) could be divided in two groups: the first one, group A, with values around $K_c = 0.4$ and $T_i = 3$ (similar values in the x-axis of the LD); and group B, with $0.5 \leq K_c \leq 0.6$

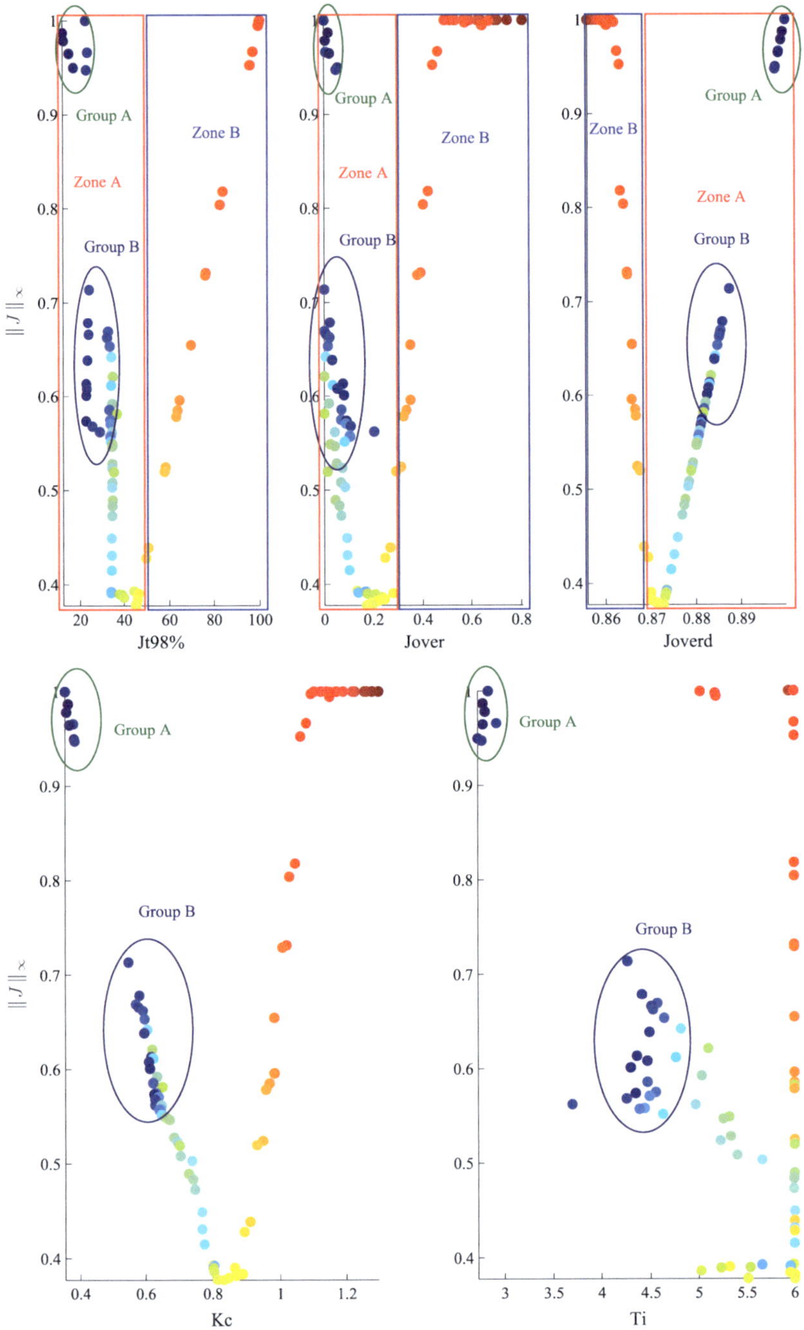

Fig. 4.2 Level diagrams for the pareto solution for $min[J_{t98\,\%}, J_{over}, J_{overd}]$ problem. $\|J\|_\infty$ used for y-axis synchronization

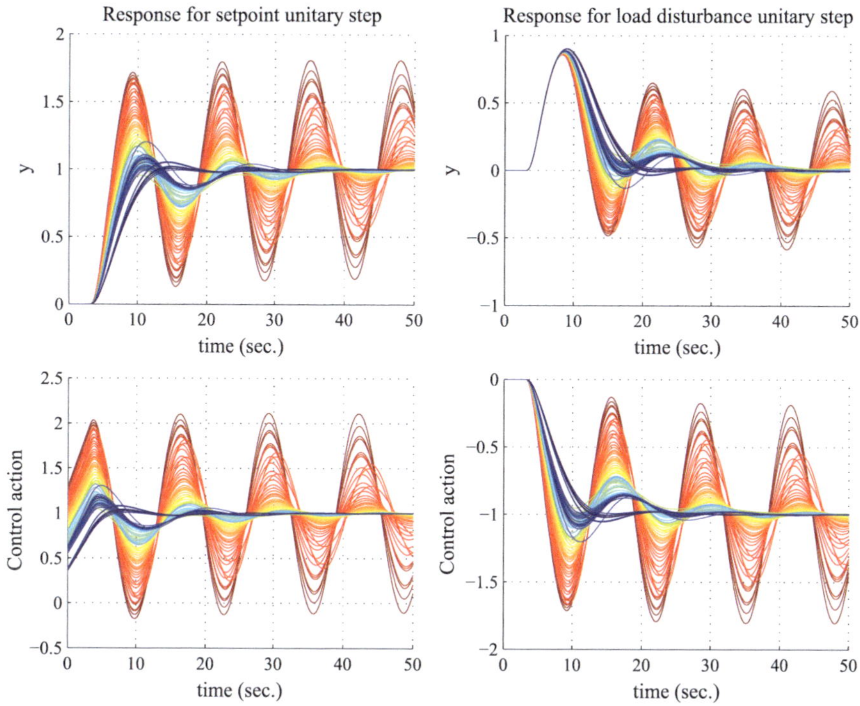

Fig. 4.3 Response for the pareto solutions for $min[J_{t98\,\%}, J_{over}, J_{overd}]$ problem

and $T_i \approx 4.5$. Remaining solutions produce high J_{over} and $J_{t98\,\%}$. Group A has slightly better performances in terms of setpoint performance (settling time under 20 s and overshoot lower than 1 %) than the group B, although the differences are not significant. However, group B has more balanced values when considering the three objectives: settling time around 20 s, overshoot around 1 % and load disturbance overshoot around 0.88.

- Notice that several Pareto Front solutions reach the upper bound of the T_i parameter (see T_i diagram). These solutions correspond to the zone B front considered as not interesting because of its poor performances in $J_{t98\,\%}$ and J_{over}. Additional optimization could be performed increasing the span of T_i, but due to the poor performances of these saturated solutions it does not probably seem some improvement.

When facing controller tuning problems it is quite useful to complement Pareto Front and Set representations with time responses for each candidate controller. Figure 4.3 shows the setpoint and load disturbance responses for the solutions obtained. Notice that most of the conclusions obtained from LD representation can be confirmed: the reduction of maximum deviation in load disturbance rejection produces important overshoots and large settling times.

Additionally, in Fig. 4.3, control action is analyzed in order to validate its feasibility for a real implementation. Huge values of control actions usually means a non-feasible implementation due to actuators limits in real applications.

Remark that performance in presence of load disturbances should have been predicted since in systems with a high delay, the PI controller cannot react immediately producing high deviation of the controlled variable. Corrections only could be produced once the controlled variable is affected by the disturbance and the result is a delayed reaction.

To avoid producing Pareto solutions with too high overshoot, several alternatives are available: apply new indicators, add constraints on overshoot or use sp-MODE-II with a pre-defined preference set to produce a more pertinent Pareto Set (this option will be explore in next chapters). Following the first option, in order to reduce simultaneously the deviation and the duration of the disturbance effect over the controlled variable, the ITAE indicator (J_{ITAEd}) will be use instead of J_{overd}. Although it is less intuitive to interpret (particular values of J_{ITAEd} and their variation are not easy to understand), it could be possible to compare solutions and to know which are better than others. Then the problem is stated as:

$$\min_{\boldsymbol{\theta}} [J_{t98\,\%}(\boldsymbol{\theta}), J_{over}(\boldsymbol{\theta}), J_{ITAEd}(\boldsymbol{\theta})] \tag{4.4}$$

$$st:$$

$$0.1 \leq K_c \leq 2$$

$$0.1 \leq T_i \leq 6$$

$$\text{Stable in closed loop}$$

The sp-MODE-II algorithm with same configuration parameters is executed and 94 Pareto solutions has been found. In Fig. 4.4 LD with ∞-norm is used for Pareto Front and Set representation. Analyzing this figure some conclusions are easy to extract:

- The ranges of $J_{t98\,\%}$ and J_{over} have been drastically reduced compared with the solutions of the previous problem. Both ranges seem more admissible than before: $J_{t98\,\%} \in [12, 24]$ s and $J_{over} \in [0, 0.09] = [0, 9]\,\%$.
- Unfortunately J_{ITAEd} is not easy to translate into a physical property, meaning it is not easy to predict which will be the time responses when load disturbances appear. Even that, the use of this objective has produced a more "pertinent" Pareto Front.
- Again, there is a conflict between setpoint response and load disturbances rejection.
- The most balanced solutions, (the ones with lower $\|\boldsymbol{J}\|_\infty$) have values under 0.4, meaning that for these solutions the worst of its objectives is under 40 % of the scale of the objective.
- The Pareto Set shows that all the solutions are quite similar $K_c \in [0.36, 0.38]$ and $T_i \approx 2.8$ s. Only a few solutions (which produce worse setpoint responses but slightly better disturbance rejections) are outside of this selection.

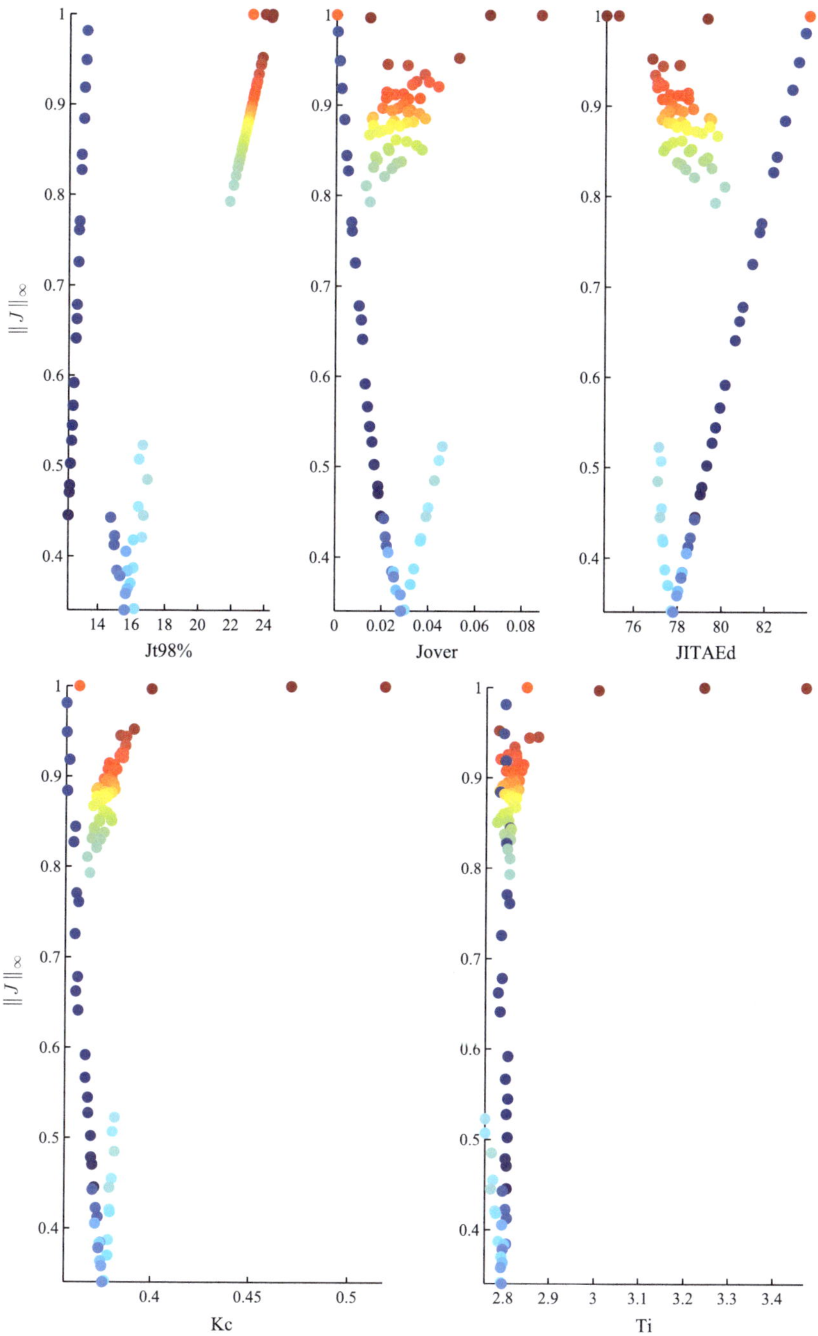

Fig. 4.4 Level diagrams for the pareto solution for $min[J_{t98\%}, J_{over}, J_{ITAEd}]$ problem. $\|J\|_\infty$ is used for y-axis synchronization

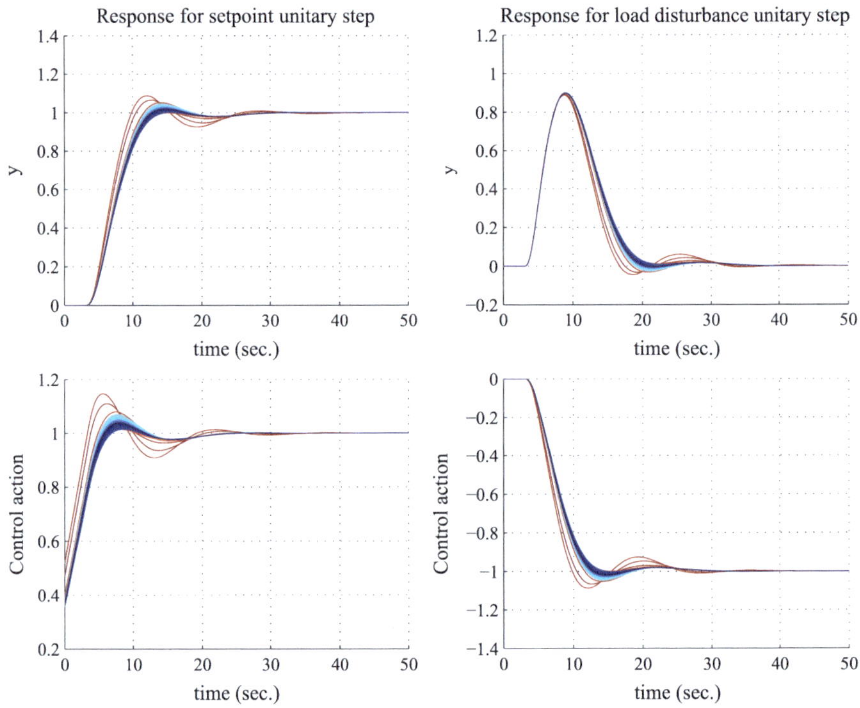

Fig. 4.5 Response for the pareto solution for $min[J_{t98\,\%}, J_{over}, J_{ITAEd}]$ problem

Figure 4.5 shows the 94 closed loop responses for setpoint and disturbance changes. The figure confirms some of the conclusions extracted from LD representation: all solutions obtained are quite similar. In this representation it is clear that it would be very difficult to obtain better performances for load disturbances rejection (not easy inspecting the values of J_{ITAEd}). All solutions reach a similar performances for load disturbance rejection and there are only some slight differences in the settling times and the oscillations produced.

To improve the reliability of the selected controller the designer could require an additional objective related to robustness. The maximum of sensitivity function (M_s) is commonly used for this purpose. Again, particular values of M_s are not easy to translate to closed loop responses with model variations but typical values are in the range of 1–2 (from more conservative/robust to more aggressive controllers).

Two approximations are analyzed: adding this indicator just for the decision making procedure or use it as a new objective $J_{Ms} = M_s$.

For the first alternative, results obtained from problem (4.4) (Fig. 4.5) are used and an additional LD axis is added with the value of J_{Ms} for the Pareto approximation. The modified LD using J_{Ms} (see Fig. 4.6) shows that almost all the solutions of Fig. 4.5

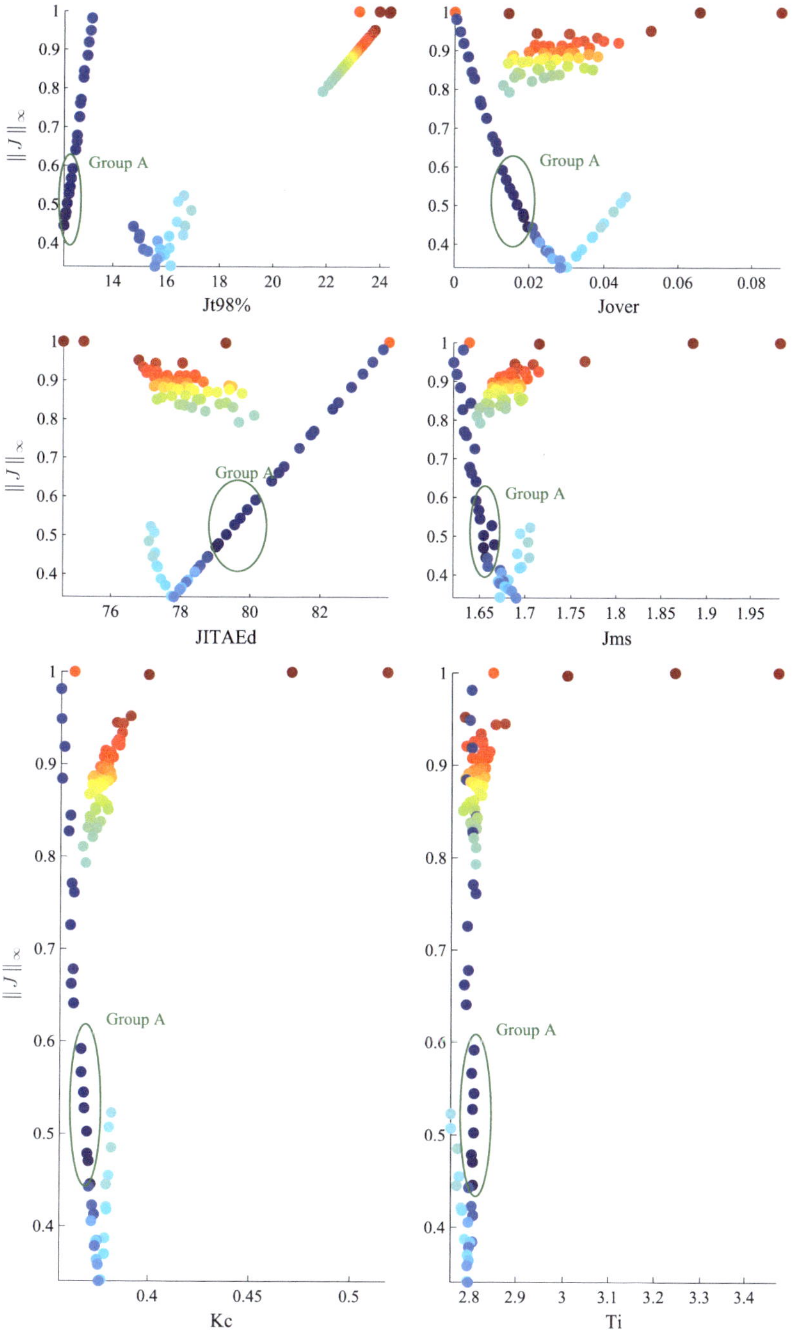

Fig. 4.6 Level diagrams for the pareto solution for $min[J_{t98\%}, J_{over}, J_{ITAEd}]$ plus an additional indicator J_{Ms}. $\|J\|_\infty$ is used for y-axis synchronization

have a $J_{Ms} \in [1.6, 1.72]$. All these values are acceptable for robustness purposes, so the selection of the final solution has to be based on the other objectives. All these solutions are in the range $K_c \in [0.36, 0.38]$ and $T_i \approx 2.8$ s.

An acceptable solution can be found inside the subset of solutions marked as Group A in Fig. 4.6. The overshoot is under 0.2, the settling time under 13 s, the *IATEd* has an average value (around 79) and the J_{Ms} indicator is around 1.65. For instance, the selected solution can be: $K_c = 0.3715$ and $T_i = 2.8057$, that gets $J_{t98\,\%} = 12.19$ s, $J_{over} = 0.0197$ %, $J_{ITAEd} = 78.7854$, and $J_{Ms} = 1.6569$.

With the second alternative the new problem (4-dimensional) is stated as:

$$\min_{\boldsymbol{\theta}}[J_{t98\,\%}(\boldsymbol{\theta}), J_{over}(\boldsymbol{\theta}), J_{ITAEd}(\boldsymbol{\theta}), J_{Ms}(\boldsymbol{\theta})] \tag{4.5}$$

$$\begin{aligned} st: & \\ & 0.1 \le K_c \le 2 \\ & 0.1 \le T_i \le 6 \\ & 1 \le J_{Ms} \le 2, \\ & \text{Stable in closed loop.} \end{aligned}$$

To avoid non-robust solutions J_{Ms} is constrained to the recommended values [1]. Again with the same parameters, sp-MODE-II algorithm offers a Pareto Set approximation of 74 points. Figure 4.7 shows the Pareto Front and Set and Fig. 4.8 shows time responses for the 74 solutions.

Analyzing the Pareto Front, it easy to see that $J_{t98\,\%}$ and J_{ITAEd} are correlated; worse $J_{t98\,\%}$ are related with worse J_{ITAEd} and vice versa. The same happens with J_{over} and J_{Ms}, but they are in contradiction with $J_{t98\,\%}$ and J_{ITAEd}. At this point, to obtain the final solution, designer likes have to be involved. Assuming that small settling times and overshoots are preferred, a tentative group of solutions should be Group A (marked as dark blue), as robustness is checked as reasonable (J_{Ms} around 1.6).

If noise sensitivity is considered, lower gains are preferred [1]. Analyzing the Pareto Front and setting the preferences in the same way than previously: $J_{t98\,\%} \le 14$ s, $J_{over} \le 0.02$, a group of possible solutions can be located (dark blue ones). Among them, the lower J_{Ms} solution is selected: $K_c = 0, 31$ and $T_i = 2, 49$ obtaining $J_{t98\,\%} = 13.95$, $J_{over} = 0.009$, $J_{ITAEd} = 85.24$ and $J_{Ms} = 1.6225$. If a lower J_{Ms} is preferred, one or several of the other objectives has to be relaxed. For instance: $K_c = 0.24$ and $T_i = 2.09$, obtaining $J_{t98\,\%} = 15.39$, $J_{over} = 0.018$, $J_{ITAEd} = 96.6$ and $J_{Ms} = 1.57$. Notice that this last solution has a lower K_c producing a better noise rejection.

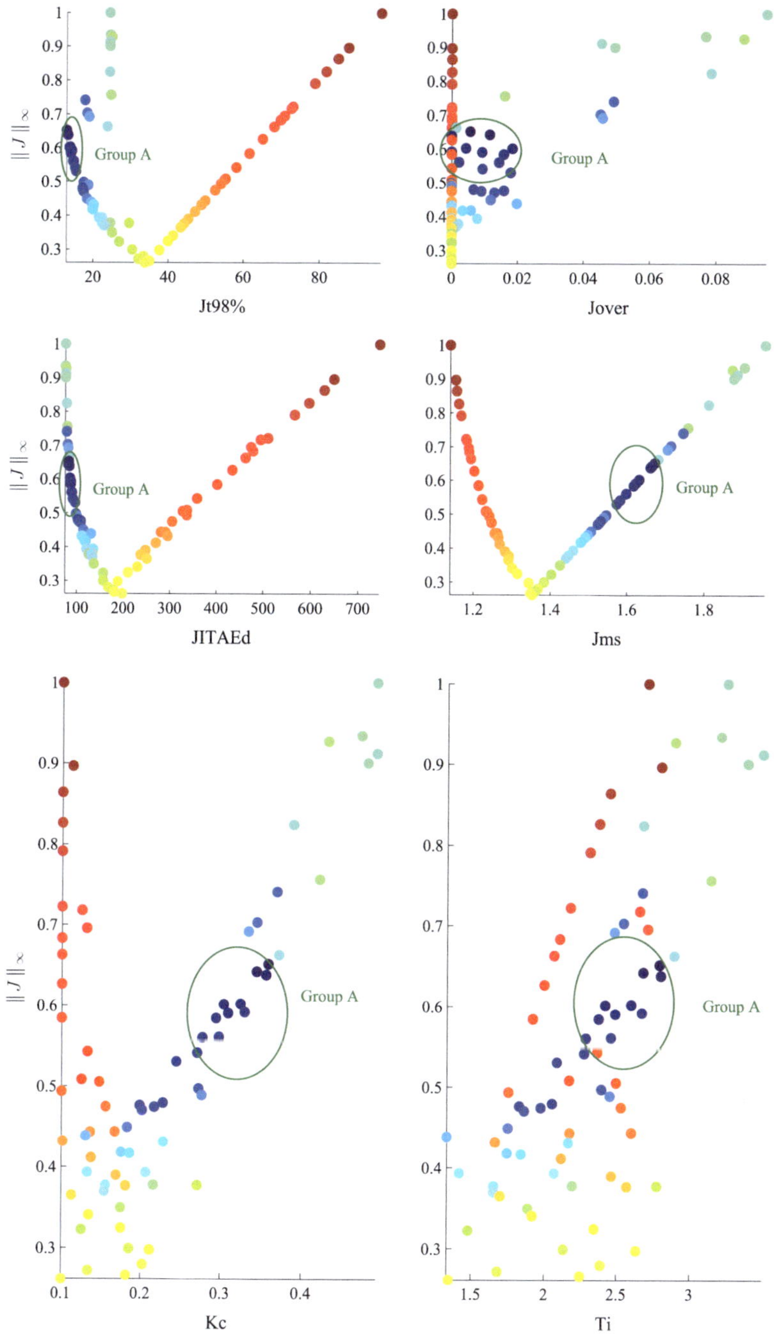

Fig. 4.7 Level diagrams for pareto solution of MOP (4.5). $\|J\|_\infty$ is used for y-axis synchronization

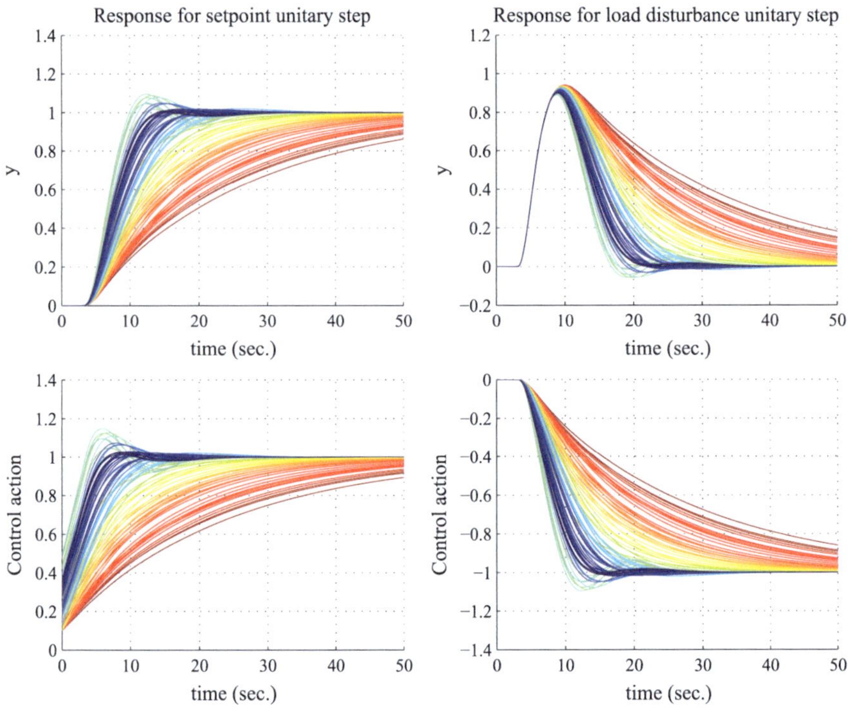

Fig. 4.8 Closed loop responses generated by pareto solution of MOP (4.5)

4.4 Performance of Some Available Tuning Rules

To evaluate some common tuning rules, firstly lets approximate the given process $P(s)$ by a FOPDT model:

$$P_a(s) = \frac{K}{Ts+1}e^{-Ls}. \tag{4.6}$$

$P_a(s)$ is approximated using the half rule defined by S. Skogestad [2]: $K = 1$, $T = 1.5$ and $L = 4.5$.

With this model, several tuning rules could be used but SIMC-rule [2] has demonstrated good performances and will be used for comparison. The SIMC-rule requires that the designer sets the desired performances by means of τ_c. When L has a high value, $\tau_c = L$ if fast response with good robustness is desired. Then the adjustment of the PI controller[1] is:

[1] For the standard PI representation.

Table 4.1 Comparison of SIMC solution and selected solution from MOOD approach

Tuning	K_c	T_i	$J_{t98\,\%}$	J_{over}	J_{ITAEd}	J_{Ms}
ZN	0.67	9.57	69.42	0	326.5	1.97
SIMC	0.17	1.50	27.43	0.049	115.7	1.59
MO approach	0.24	2.09	15.39	0.018	96.6	1.57

$$K_c = \frac{T}{K(\tau_c + L)} = 0.17, \tag{4.7}$$

$$T_i = \min\{T, 4(\tau_c + L)\} = 1.5. \tag{4.8}$$

Additionally the well known Ziegler-Nichols [3] tuning method is also compared. The "ultimate" gain and period can be computed from the process model giving $K_u = 1.48$ and $P_u = 11.49$. The controller parameters results in:

$$K_c = 0.45 \cdot K_u = 0.67, \tag{4.9}$$

$$T_i = P_u/1.2 = 9.57. \tag{4.10}$$

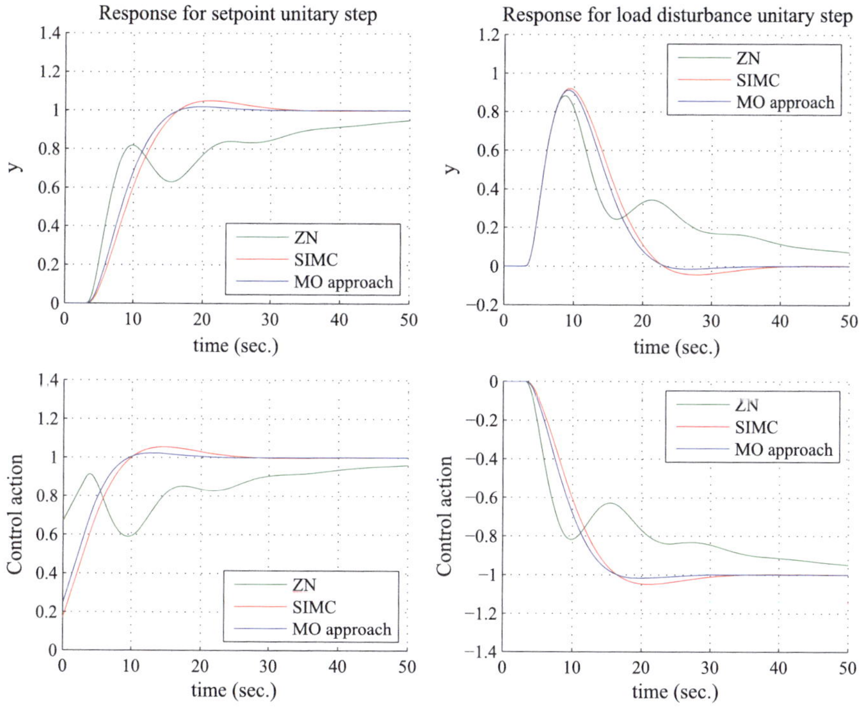

Fig. 4.9 Response for the PI tuned with Ziegler Nichols and SIMC-rule versus the selected from MOOD approach

The objective values for these particular solutions are shown in Table 4.1 and the responses are shown in Fig. 4.9. Remark that the resulting Ziegler-Nichols T_i is outside the bounds established in the MOP. Clearly MO approach and SIMC solutions offers better behavior than ZN (both dominate the ZN solution, since it is not a Pareto solution). Although SIMC solution has good performances it is also outside the Pareto Front obtained from MOP (4.5) and the solution selected from the MOOD approach is better than SIMC in the three objectives (dominated solution).

4.5 Conclusions

This simple example for SISO controller tuning has shown the basic steps of the MOOD approach. It is supposed that the designer wants a better understanding of the trade-off among the different objectives and wants to know the limitations of the different available controllers. For this purpose multiobjective methodology is worthwhile.

The process model used in the example has three poles, an important delay and a load disturbance. The controller to tune is a PI, no other control structure has been evaluated and the comparison between different alternatives is out of the scope of this example. In further examples, concept design (different control alternatives) will be introduced in the MOOD approach. The aim of this example has been, exclusively, to obtain the best control tuning for a particular controller considering the control engineer preferences.

One of the topics pointed up has been the importance of the objectives used to attain designer requirements. MOOD requires a high participation of the designer: setting preferences and analyzing the Pareto solutions. This implies that the designer has to be able to interpret accurately the values of the different objectives in such a way that she/he could select the best solution according to her/his preferences. In this example, for setpoint response, two intuitive indicators have been selected as objectives: settling time and overshoot. With these two indicators, the designer could predict the shape of the time response and easily understand if a particular solution is close to his preferences.

For load disturbances rejection, a first attempt was using the maximal deviation of the controlled variable when a unitary step disturbance is applied. This indicator is easy to interpret but the obtained results are not satisfactory since most of the solutions are out of the area of interest. That suggests there is room for new research in intuitive but useful objectives that better gather the designer preferences. In fact, the designer always requires a 'pertinent' set of solutions. The capabilities of pertinency developed in sp-MODE-II have not been used in this example (they will be exploited in further examples), but even if the front is pertinent according to the designer preferences it could be usefulness due to an inappropriate objective selection.

After, an alternative indicator, ITAE, has been used. It is not so intuitive, meaning that the designer cannot predict the load disturbances rejection behavior uniquely with the particular value of ITAE. It only offers the possibility of comparing different values looking for lower values (it is supposed that a lower value means a better behavior). With this indicator used as objective, the set of solutions are closer to designer preferences and could be used for analyzing the performance of the different alternatives.

Finally, an additional objective has been added in order to consider robustness of the controller: maximum of the sensibility function (M_s) which is a common and useful indicator for that purpose. But again, the indicator is not intuitive, it is not possible to predict what will be the model uncertainty robustness behavior looking at a particular value of Ms, but it is useful for comparison between different solutions.

As a final remark, it is important to point out the contribution of the graphical tools for the Pareto Front and Set analysis. The LD representation is used for this purpose and, although it requires an initial training to understand better what type of information it supplies, it has proved to be a good tool for high dimensional set analysis. It is undeniable that for control tuning, complementary graphs can be very useful, in fact, time responses showing not only the controlled variables but also the manipulated ones should be presented together with LD.

Finally, other well-known tuning techniques has been presented and compared with the solution obtained from MOOD methodology.

References

1. Åström K, Hägglund T (1995) PID controllers: theory, design and tuning. Systems, and Automation Society, ISA - The Instrumentation
2. Skogestad S (2003) Simple analytic rules for model reduction and PID controller tuning. J Process Control 13(4):291–309
3. Ziegler J, Nichols N (1942) Optimum settings for automatic controllers. ASME 64:759–768

Chapter 5
Controller Tuning for Multivariable Processes

Abstract In this chapter, a multivariable controller is tuned by means of a multiobjective optimization design procedure. For this design problem, several specifications are given, regarding individual control loops and overall performance. Due to this fact, a many-objectives optimization problem is stated. In such problems, algorithms could face problems due to the dimensionality of the problem, since their mechanisms to improve convergence and diversity may conflict. Therefore, some guidelines to deal with this optimization process are commented. The aforementioned procedure will be used to tune a multivariable PI controller for the well known Wood and Berry distillation column process using different algorithms.

5.1 Introduction

So far, we have been dealing with single input and a single output (SISO) processes. Nevertheless, a wide variety of industrial processes are multivariable, that is, with multiple inputs and multiple outputs (MIMO). In such instances, the controller tuning task could be more defiant, since coupling effects and interacting dynamics have to be taken into account by the designer.

MIMO processes are quite common in industry and several control techniques have been used for such processes like predictive control and state space feedback techniques. Nonetheless PI like controllers remain as a preferred choice for the lower control layer, due to its simplicity; given that in industrial environments is common to dealt with hundreds of control loops, using a simple controller structure wherever is possible, alleviates the control engineer's work, and allows us to focus on more complex (or sensitive) control loops.

In order to show the usability of the MOOD procedure for a MIMO process, Wood and Berry distillation column control problem [1, 13] will be used. It is a classical benchmark for multivariable control, which describes the dynamics of overhead and bottom composition of methanol and water in the column. It is a popular MIMO process where several control techniques have been evaluated as well as controller tuning using evolutionary algorithms [3–5, 7] and evolutionary multiobjective optimization [9].

G. Reynoso Meza et al., *Controller Tuning with Evolutionary Multiobjective Optimization*, Intelligent Systems, Control and Automation: Science and Engineering 85, DOI 10.1007/978-3-319-41301-3_5

5.2 Model Description and Control Problem

The well-known distillation column model defined by Wood and Berry will be used [1, 13] (see Fig. 5.1). For binary distillation, composition of methanol and water of overhead product X_D [%] and bottom product X_B [%] has to be controlled by means of the reflux and steam flows R [lbs/min], S [lbs/min] respectively. Typical steady state operating conditions are $X_D = 96$ %, $X_B = 0.5$ %, $R = 1.95$ lbs/min and $S = 1.71$ lbs/min. For this equilibrium point the multivariable process is modelled as:

$$
\begin{aligned}
Y(S) = \quad & P(S) \quad U(s) \quad + \quad D(s) \quad N(S) \quad = \\
\begin{bmatrix} X_D(s) \\ X_B(s) \end{bmatrix} = & \begin{bmatrix} P_{11}(s) & P_{12}(s) \\ P_{21}(s) & P_{22}(s) \end{bmatrix} \begin{bmatrix} R(s) \\ S(s) \end{bmatrix} + \begin{bmatrix} D_{11}(s) & D_{12}(s) \\ D_{21}(s) & D_{22}(s) \end{bmatrix} \begin{bmatrix} F(s) \\ X_F(s) \end{bmatrix} \\
= & \begin{bmatrix} \frac{12.8e^{-s}}{16.7s+1} & \frac{-18.9e^{-3s}}{21s+1} \\ \frac{6.6e^{-7s}}{10.9s+1} & \frac{-19.4e^{-3s}}{14.4s+1} \end{bmatrix} \begin{bmatrix} R(s) \\ S(s) \end{bmatrix} + \begin{bmatrix} \frac{3.8e^{-8.1s}}{14.9s+1} & \frac{0.22e^{-7.7s}}{22.8s+1} \\ \frac{4.9e^{-3.4s}}{13.2s+1} & \frac{0.14e^{-9.2s}}{12.1s+1} \end{bmatrix} \begin{bmatrix} F(s) \\ X_F(s) \end{bmatrix}.
\end{aligned} \tag{5.1}
$$

Such process will be controlled with a decentralized PI controller structure $C(s)$ (see Fig. 5.2):

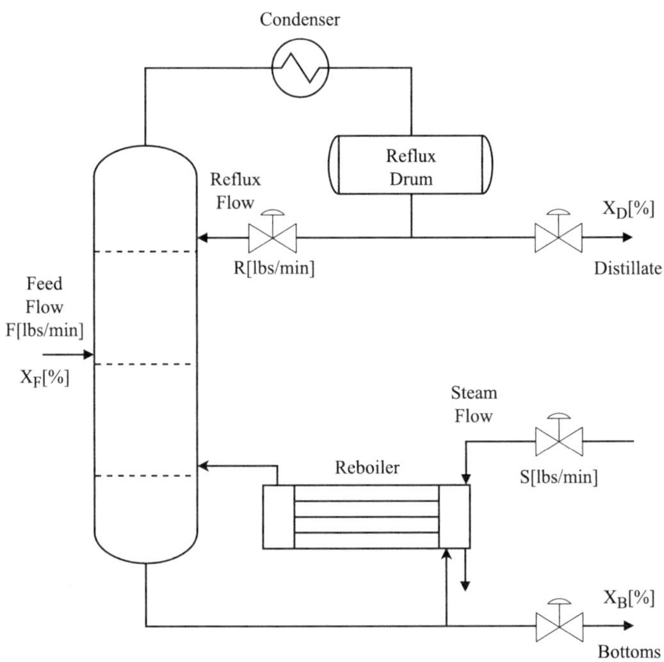

Fig. 5.1 Process flow diagram for Wood and Berry distillation column

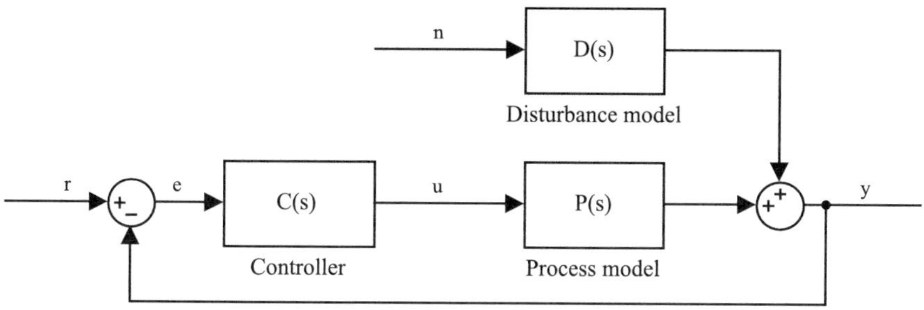

Fig. 5.2 Basic loop for descentralized PI controller

$$C(s) = \begin{bmatrix} Kc_1\left(1 + \frac{1}{Ti_1 s}\right) & 0 \\ 0 & Kc_2\left(1 + \frac{1}{Ti_2 s}\right) \end{bmatrix}. \tag{5.2}$$

The main task of the control loop is to reject disturbances due to changes on the column feed flow F [lb/mins] and its composition X_F [%].

5.3 The MOOD Approach

A MOP statement with 4 design objectives will be stated. The first two related with performance of individual control loops; for such purpose we use the IAE index (Eq. 2.9) for the overhead and bottom products:

$$J_{IAE_{X_D}}(\boldsymbol{\theta}) \ \ [\% \cdot min], \tag{5.3}$$

$$J_{IAE_{X_B}}(\boldsymbol{\theta}) \ \ [\% \cdot min]. \tag{5.4}$$

Since design objectives are related with the control action and robustness of individual loops the TV index (see Eq. 2.18) for the reflux and steam flows, R and S, will be used:

$$J_{TV_R}(\boldsymbol{\theta}) \ \ [lbs/min], \tag{5.5}$$

$$J_{TV_S}(\boldsymbol{\theta}) \ \ [lbs/min]. \tag{5.6}$$

Both objectives will be evaluated through simulating a load disturbance n in the process. Decision variables $\boldsymbol{\theta}$ are the proportional gain and integral time [min] for each control loop $\boldsymbol{\theta} = [Kc_1, Ti_1, Kc_2, Ti_2]$. As commented in Chap. 1, a MOOD procedure for controller tuning is valuable when:

- *It is difficult to find a controller with a reasonable balance among design objectives* or,
- *It is worthwhile analyzing the trade-off among controllers (design alternatives).*

For the case of PI controller tuning, there are several tuning rules available for designers, for SISO and MIMO processes. Therefore, the added value of a MOOD procedure in such instances is related with the above commented difficulty. Since most of the time, a tuning rule will be available for a control loop, it could be expected that the MOOD procedure will bring a set of solutions better than such tuning rule; that is, the designer has a tuning rule or procedure that may be used as a reference case, and therefore it is expected to approximate a Pareto Front which dominated its overall performance. Let's denote hereafter such tuning rule as the *reference tuning rule*. Examples of such tuning rules could be:

- *A manual procedure performed by an experienced engineer.*
- *A well established and formal tuning procedure developed for specific controllers under specific circumstances, which applies to the current problem.*
- *The performance of a complex controller, which is stated as reference for a less complex structure.*

For the sake of simplicity of our example, the *reference tuning rule* applied is the one specified by the Biggest Log Modulus Tuning (BLT) criterion for diagonal PI controllers in MIMO processes [6]. The BLT criterion is a well known tuning rule, widely used to evaluate different controller performances in the scientific literature. The BLT criterion proposes a *de-tuning* of the proportional gains of the controllers obtained by the Ziegler-Nichols method for each individual loop, in order to fulfill a maximum value of the closed loop log modulus L_{cm}:

$$\begin{aligned} L_{cm} &= 20 \log \left| \frac{W(s)}{1 + W(s)} \right|, \\ W(s) &= -1 + \det\left(I + P(s)C(s)\right). \end{aligned} \tag{5.7}$$

It has been suggested that an empirical value of an NxN multivariable process is $L_{cm}^{max} = 2N$ [6]. Then, applying BLT criterion to control the process (5.1), the following tuning parameters are obtained: $Kc_1 = 0.375, Ti_1 = 8.29, Kc_2 = -0.0750$ and $Ti_2 = 23.6$. Therefore, the reference solution $\boldsymbol{\theta}_R$ specified by the *reference tuning rule* is $\boldsymbol{\theta}_R = [0.375, 8.290, -0.075, 23.600]$ and the designer is interested in improving its performance $J(\boldsymbol{\theta}_R)$ (that is $\boldsymbol{\theta} \preceq \boldsymbol{\theta}_R$) or finding another suitable controller in its *surroundings* on the objective space (that is, $J_i(\boldsymbol{\theta}) < J_i(\boldsymbol{\theta}_R) + \Delta_{J_i}, i \in [1, \cdots, m]$).

Three different instances, detailed in Chap. 3, will be evaluated:

1. A simple DE algorithm [2, 11, 12], using a preference matrix to minimize the GPP index.
2. The sp-MODE algorithm [10], with a basic pertinency improvement with bound on design objectives.

Table 5.1 Parameters used for DE, sp-MODE and sp-MODE-II. Further details in Chap. 3

			Parameter	Value
sp-MODE-II	**sp-MODE**	**DE**	**Evolutionary mechanism**	
			F (Scaling factor)	0.5
			Cr (Crossover rate)	0.9
			N_p (Population)	50
			Pruning mechanism	
			$\boldsymbol{\beta}^{\epsilon}$ (Arcs)	40
			Pertinency mechanism	
			J_{GPP}^{max}	$J_{GPP}([1.0, 1.0, 1.1, 1.1])$
			$\overline{car(J_P^*)}$	40

3. The sp-MODE-II algorithm [8], with the full set of preferences.

In each case, parameters used for optimization are shown in Table 5.1 (accordingly with guidelines in Chap. 3). While normally the designer will choose a desired approach to tackle the optimization problem at hand, we will evaluate three instances in order to show their structural differences to approximate a Pareto Front and to bring a useful set of solutions to the decision maker.

To improve the understanding of objectives, and in order to build the preference matrix, objectives will be normalized respect to $\boldsymbol{J}(\boldsymbol{\theta}_R)$. This will facilitate the visualization since the improvements over design objectives for the reference case will be more evident. Consequently, the MOP can be stated as:

$$\min_{\boldsymbol{\theta}} \boldsymbol{J}(\boldsymbol{\theta}) = [\hat{J}_{IAE_{X_D}}(\boldsymbol{\theta}), \hat{J}_{IAE_{X_B}}(\boldsymbol{\theta}), \hat{J}_{TV_R}(\boldsymbol{\theta}), \hat{J}_{TV_S}(\boldsymbol{\theta})] \tag{5.8}$$

where

$$\boldsymbol{\theta} = [Kc_1, Kc_2, Ti_1, Ti_2] \tag{5.9}$$

subject to:

$$\begin{aligned} 0 \leq Kc_1 &\leq 1 \\ -1 \leq Kc_2 &\leq 0 \\ 0 < Ti_{1,2} &\leq 50 \\ L_{cm}(\boldsymbol{\theta}) &< 4 \end{aligned} \tag{5.10}$$

and

$$\boldsymbol{\theta}_R = [0.375, 8.290, -0.075, 23.600] \tag{5.11}$$

$$\hat{J}_{IAE_{X_D}}(\boldsymbol{\theta}) = \frac{J_{IAE_{X_D}}(\boldsymbol{\theta})}{J_{IAE_{X_D}}(\boldsymbol{\theta}_R)} = \frac{J_{IAE_{X_D}}(\boldsymbol{\theta})}{2.6} \tag{5.12}$$

$$\hat{J}_{IAE_{X_B}}(\boldsymbol{\theta}) = \frac{J_{IAE_{X_B}}(\boldsymbol{\theta})}{J_{IAE_{X_B}}(\boldsymbol{\theta}_R)} = \frac{J_{IAE_{X_B}}(\boldsymbol{\theta})}{32} \tag{5.13}$$

$$\hat{J}_{TV_R}(\boldsymbol{\theta}) = \frac{J_{TV_R}(\boldsymbol{\theta})}{J_{TV_R}(\boldsymbol{\theta}_R)} = \frac{J_{TV_R}(\boldsymbol{\theta})}{0.19} \tag{5.14}$$

$$\hat{J}_{TV_S}(\boldsymbol{\theta}) = \frac{J_{TV_S}(\boldsymbol{\theta})}{J_{TV_S}(\boldsymbol{\theta}_R)} = \frac{J_{TV_S}(\boldsymbol{\theta})}{0.12}. \tag{5.15}$$

The last constraint ($L_{cm}(\boldsymbol{\theta}) < 4$) ensures a fair comparison with the *reference tuning rule*, as well as overall robustness. Indexes will be calculated with time responses obtained from closed loop simulations when a step change of 0.34 lb/min on the feed flow F is applied. This is justifiable since the most important changes on the systems are due to feed flow changes.

Some extra constraints are added to (5.10) when sp-MODE algorithm is used in order to add a basic pertinence mechanism:

$$\begin{aligned} \hat{J}_{IAE_{X_D}}(\boldsymbol{\theta}) &< 1 \\ \hat{J}_{IAE_{X_B}}(\boldsymbol{\theta}) &< 1 \\ \hat{J}_{TV_R}(\boldsymbol{\theta}) &< 1 + 10\,\% \\ \hat{J}_{TV_S}(\boldsymbol{\theta}) &< 1 + 10\,\%. \end{aligned} \tag{5.16}$$

The preference set for sp-MODE-II algorithm is shown in Table 5.2. Same preferences will be considered when DE algorithm is used to minimize the GPP index. Notice that, the T_Vector is defined as $\boldsymbol{J}^T = [1.0, 1.0, 1.1, 1.1]$ meaning that the

Table 5.2 Preference matrix $\mathfrak{P}$ for multivariable PI controller tuning. Five preference ranges have been defined: *Highly Desirable* (HD), *Desirable* (D), *Tolerable* (T) *Undesirable* (U) and *Highly Undesirable* (HU)

Preference matrix $\mathfrak{P}$						
Objective	← HD →	← D →	← T →	← U →	← HU →	
	J_i^0	J_i^1	J_i^2	J_i^3	J_i^4	J_i^5
$\hat{J}_{IAE_{X_D}}(\boldsymbol{\theta})$	0.70	0.80	0.90	1.00	2.00	5.00
$\hat{J}_{IAE_{X_B}}(\boldsymbol{\theta})$	0.20	0.30	0.50	1.00	2.00	5.00
$\hat{J}_{TV_R}(\boldsymbol{\theta})$	0.80	0.90	1.00	1.10	2.00	5.00
$\hat{J}_{TV_S}(\boldsymbol{\theta})$	0.80	0.90	1.00	1.10	2.00	5.00

designer is willing to tolerate controllers that might use 10 % more control action than the reference controller $\boldsymbol{\theta}_R$, but not lesser performance. Regarding the D_Vector, $\boldsymbol{J}^D = [0.9, 0.5, 1.0, 1.0]$, the designer is seeking to improve firstly the control of X_B [%] and that desirable solutions will be those that, using the same control effort measure, achieve a better performance than the reference controller $\boldsymbol{\theta}_R$.

With the first approach (DE algorithm), a single solution $\boldsymbol{\theta}_{GPP}$ is calculated where

$$\boldsymbol{\theta}_{GPP} = [0.459, 9.818, -0.067, 6.986]$$

The sp-MODE approach approximated a Pareto Front with 1439 solutions (see Level Diagrams in Figs. 5.3a and 5.4a), while the sp-MODE-II reached one with 29 solutions (Figs. 5.3b and 5.4b). Focusing firstly in the Pareto Front approximation, it is evident that the compactness of $\boldsymbol{\Theta}^*_{P_2}$, $\boldsymbol{J}^*_{P_2}$ are approximations from the sp-MODE-II versus the spreading and covering of approximations $\boldsymbol{\Theta}^*_{P_1}$, $\boldsymbol{J}^*_{P_1}$ from sp-MODE. In the former case, the DM needs to concentrate its analysis in a more manageable set of solutions. In the latter, it is possible to fully appreciate the trade-off exchange through all the Pareto front. That is, the former is more useful for the analysis and selection of a preferable solution, while the latter could offer a better perspective of the overall trade-off, and could be helpful to have a better understanding of the control problem and its trade-off between performance and control cost.

The natural questions here are:

- *why not actively seek for the solution $\boldsymbol{\theta}_{GPP}$ with the lowest GPP norm, as in the case of the DE algorithm? or*
- *why not choose directly the solution with the lowest GPP norm from the $\boldsymbol{\Theta}^*_{P_1}$ approximation provided by the sp-MODE algorithm?*

In the first instance, while a solution minimizing the GPP index will bring the most preferable solution according to a preference set matrix, it gives no idea about the trade-off in the surroundings of this preferable solution and perhaps the DM may prefer other solutions in that area with a more reasonable trade-off for the problem at hand. This can be done only via analyzing the Pareto Front approximation. In the second instance, it could be worthwhile having a semi-automatic procedure to select a solution from $\boldsymbol{\Theta}^*_{P_1}$, $\boldsymbol{J}^*_{P_1}$; nevertheless, again, the DM may prefer other solutions in the surroundings, seeking for a more convenient trade-off.

In that sense, a practical approximation $\boldsymbol{\Theta}^*_{P_2}$ (giving $\boldsymbol{J}^*_{P_2}$) can be built with sp-MODE-II, which focus in a compact set of solutions covering the most preferable region of the Pareto Front. According to this, the sp-MODE-II approach is an in-between alternative for a full Pareto Front approximation (sp-MODE) and a single solution (DE+GPP). Again, it depends on the DM needs. If the designer needs a full knowledge of the problem, she/he may prefer a sp-MODE-like option. If there is a need to focus the designer's attention in the most preferable region and select a solution, an sp-MODE-II-like option could be more practical. However, if the DM is comfortable and confident with the preference matrix and needs a solution, then a DE-like approach should be used.

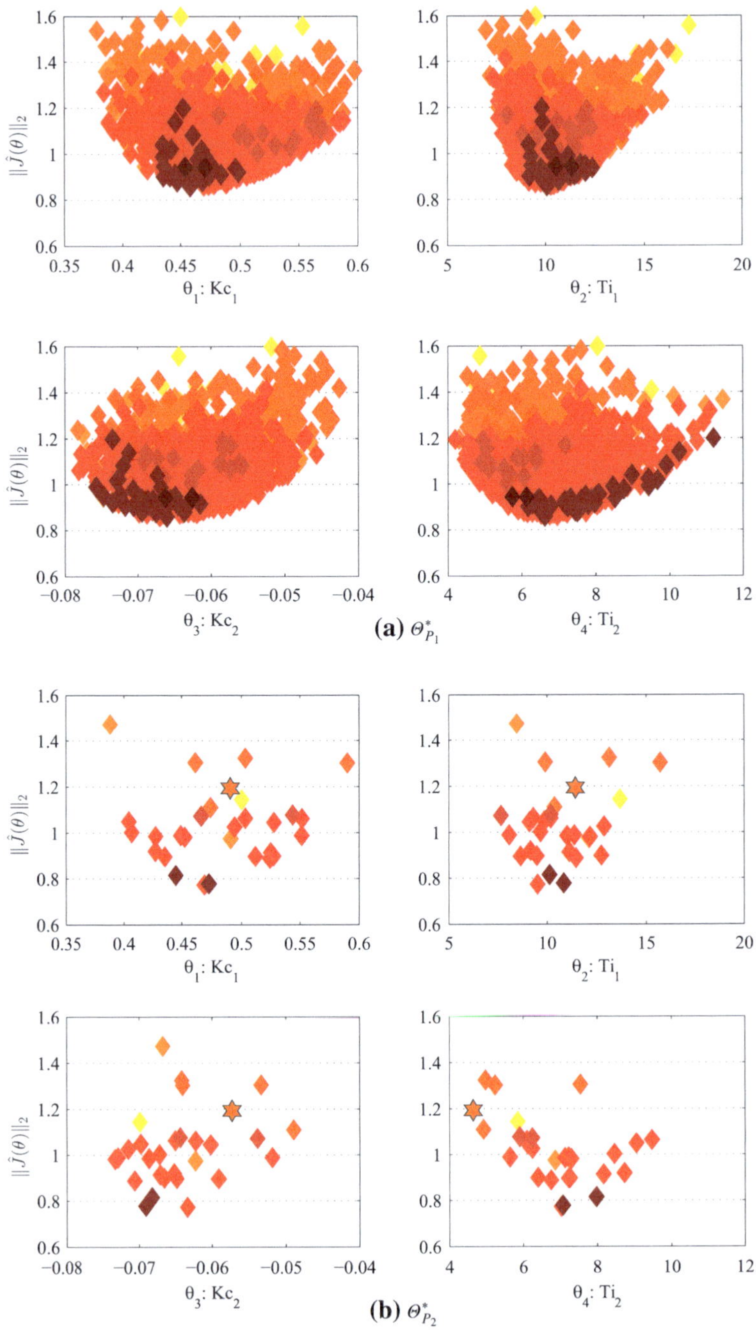

Fig. 5.3 Pareto set approximated by **a** sp-MODE ($\boldsymbol{\Theta}^*_{P_1}$) and **b** sp-MODE-II ($\boldsymbol{\Theta}^*_{P_2}$)

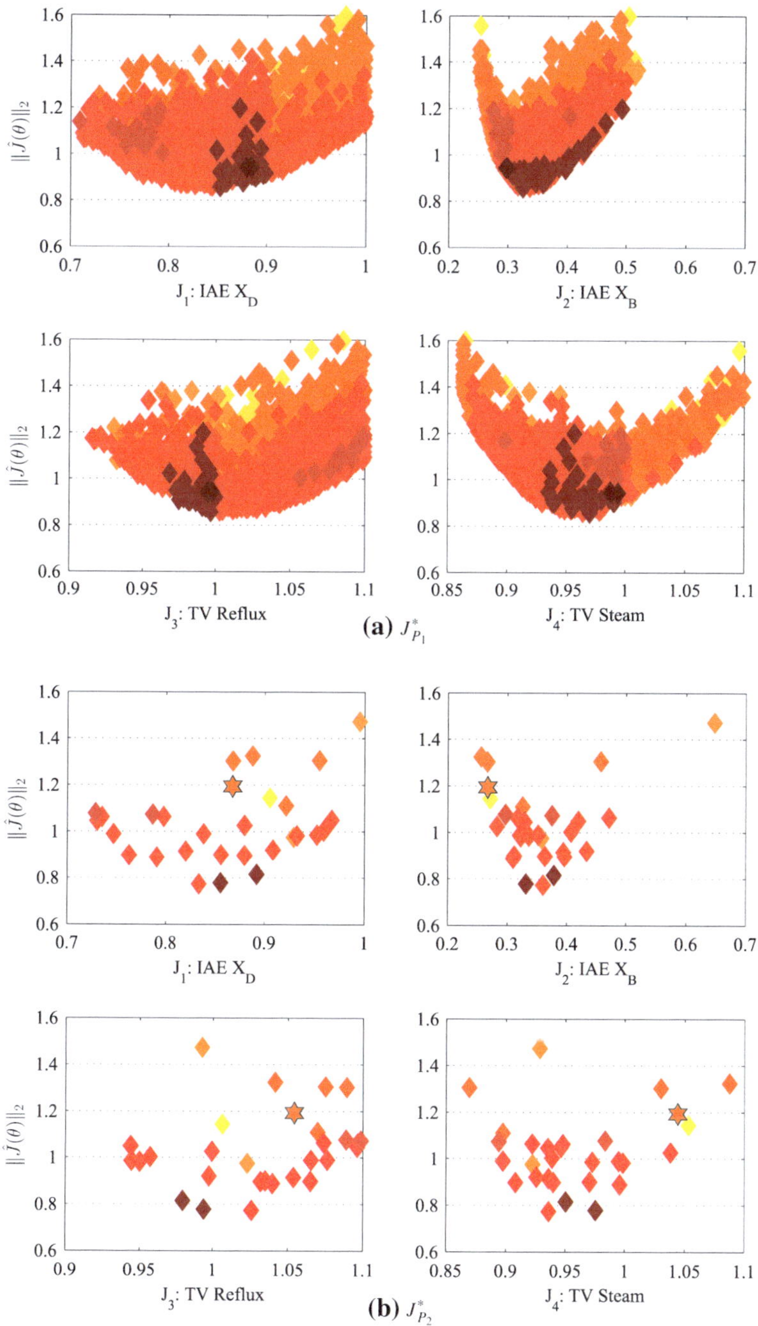

Fig. 5.4 Pareto front approximated by **a** sp-MODE ($\boldsymbol{J}^*_{P_1}$) and **b** sp-MODE-II ($\boldsymbol{J}^*_{P_2}$)

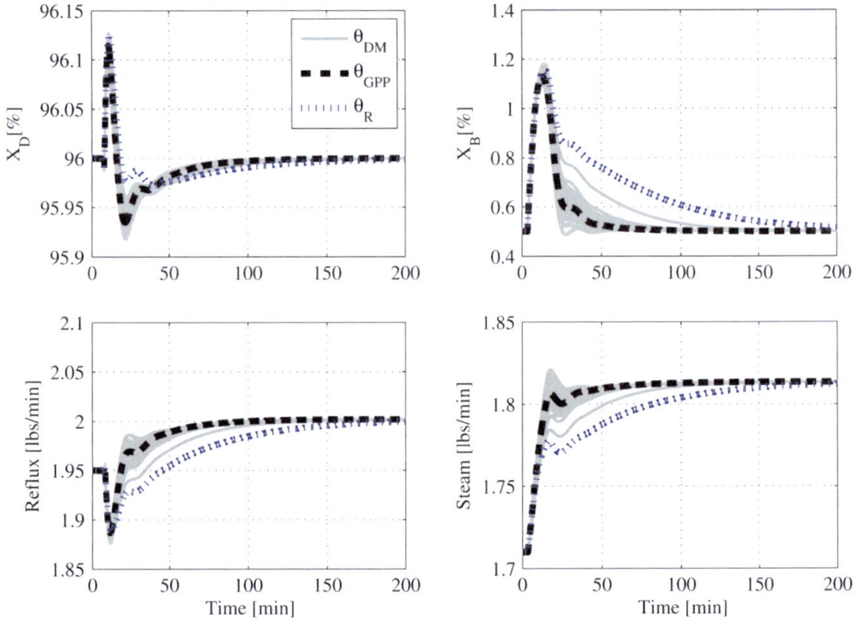

Fig. 5.5 Time response comparison for a change in the feed flow F of 0.34 lb/min (optimization test)

Table 5.3 Performance X_D for a change of 0.34 lb/min in the feed flow F (optimization test)

	Overshoot	$t_{98\,\%}$
$\boldsymbol{\Theta}^*_{P_2}$	[0.10, 0.13]	[14.00, 29.20]
$\boldsymbol{\theta}_{GPP}$	0.12	25.40
$\boldsymbol{\theta}_R$	0.12	15.90

Table 5.4 Performance X_B for a change of 0.34 lb/min in the feed flow F (optimization test)

	Overshoot	$t_{98\,\%}$
$\boldsymbol{\Theta}^*_{P_2}$	[0.61, 0.68]	[23.00, 86.80]
$\boldsymbol{\theta}_{GPP}$	0.63	36.60
$\boldsymbol{\theta}_R$	0.65	139.30

Finally in Fig. 5.5 closed loop time responses of the reference controller $\boldsymbol{\theta}_R$, solution from DE approach $\boldsymbol{\theta}_{GPP}$ and solutions from $\boldsymbol{\Theta}^*_{P_2}$ are compared. The same time response test used for optimizations is depicted. Additional performance indicators are shown in Tables 5.3 and 5.4. It can be notice that the $\boldsymbol{\theta}_{GPP}$ solution is, in the majority of the indicators, better than the $\boldsymbol{\theta}_R$ solution; nevertheless, it sacrifices the performance on the settling time of the upper product (around 60 %) in order to improve the performance of the settling time in the bottom product (around 74 %). That is, there is an exchange between settling time performance between individual loops. In the case of the solutions from $\boldsymbol{\Theta}^*_{P_2}$, intervals for each indicator are shown. In all cases $\boldsymbol{\theta}_{GPP}$ lies on such intervals; as expected, since the pruning mechanism in the sp-MODE-II algorithm uses the same preference matrix as the DE approach (in fact, the DE solution might be contained in the Pareto Set approximated by sp-MODE-II).

After analyzing the $\boldsymbol{J}^*_{P2}$ approximation, a solution $\boldsymbol{\theta}_{DM} \in \boldsymbol{\Theta}^*_{P2}$ is selected (depicted with a ★):

$$\boldsymbol{\theta}_{DM} = [0.490, 11.436, -0.057, 4.645]$$

which is basically a solution with better performance in $\hat{J}_{TV_R}(\boldsymbol{\theta})$ and $\hat{J}_{TV_S}(\boldsymbol{\theta})$ than a solution with better GPP index. Now, further control tests will be performed with $\boldsymbol{\theta}_R$, $\boldsymbol{\theta}_{GPP}$ and $\boldsymbol{\theta}_{DM}$.

5.4 Control Tests

The reference solution $\boldsymbol{\theta}_R$, the solution with the lowest GPP, $\boldsymbol{\theta}_{GPP}$, and the solution selected through an analysis of the sp-MODE-II Pareto Front, $\boldsymbol{\theta}_{DM}$ will undergo further evaluation. This follows the idea that, even when we have used a specific control test in order to seek for a controller with a preferable trade-off, it might have a different behavior on different circumstances.

By this reason, three different tests are analyzed:

1. Closed loop response for a step change of −0.5 % in the feed flow composition X_F (Fig. 5.6 and Tables 5.5, 5.6 show results from such test).
2. Closed loop response for a setpoint step change from 0.5 to 0.75 % in the bottom composition X_B (Fig. 5.7 and Tables 5.7, 5.8 show results from such test).
3. Closed loop response for a setpoint step change from 96.0 to 95.5 % in the overhead composition X_D (Fig. 5.8 and Tables 5.9, 5.10 show results from such test).

As it can be appreciated, $\boldsymbol{\theta}_{GPP}$, $\boldsymbol{\theta}_{DM}$ have better overall performance compared with the $\boldsymbol{\theta}_R$ controller. Main differences between $\boldsymbol{\theta}_{GPP}$, $\boldsymbol{\theta}_{DM}$ are appreciated in Test 1 and 2, where the latter sacrifices IAE performance in order to get better settling time.

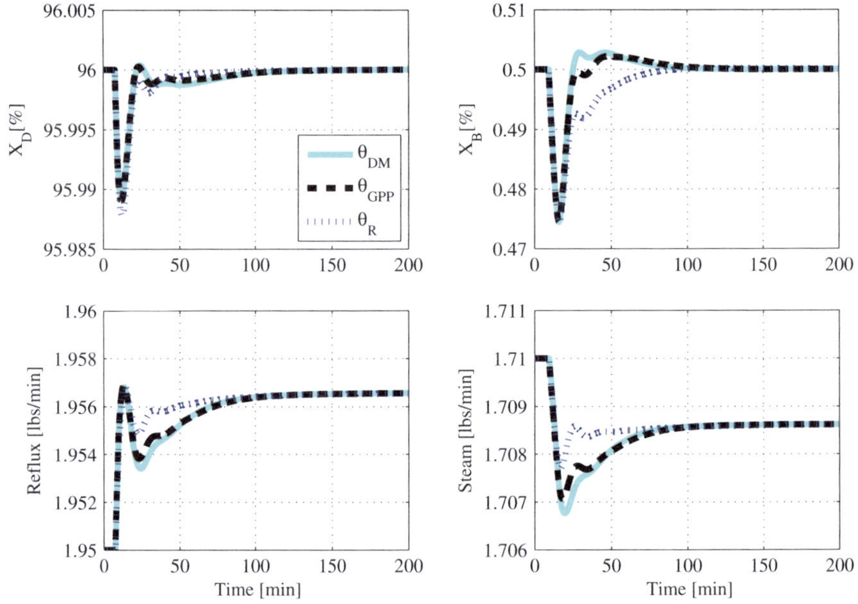

Fig. 5.6 Time response comparison for a change in the feed flow composition X_F of -0.5 % (Test 1)

Table 5.5 Performance X_D for a change in the feed flow composition X_F of -0.5 % (Test 1)

	IAE	ITAE	ISE	ITSE	Overshoot	$t_{98\,\%}$
θ_{DM}	1.55	55.96	$8e-3$	0.12	$4e-4$	105.91
θ_{GPP}	1.41	47.31	$7e-3$	0.11	$2e-4$	103.55
θ_R	1.45	36.21	$9e-3$	0.13	0.00	71.56

Table 5.6 Performance X_B for a change in the feed flow composition X_F of -0.5 % (Test 1)

	IAE	ITAE	ISE	ITSE	Overshoot	$t_{98\,\%}$
θ_{DM}	3.40	98.55	0.05	0.79	0.57	82.80
θ_{GPP}	3.33	105.46	0.04	0.79	0.43	95.91
θ_R	4.69	147.66	0.06	1.21	0.04	76.85

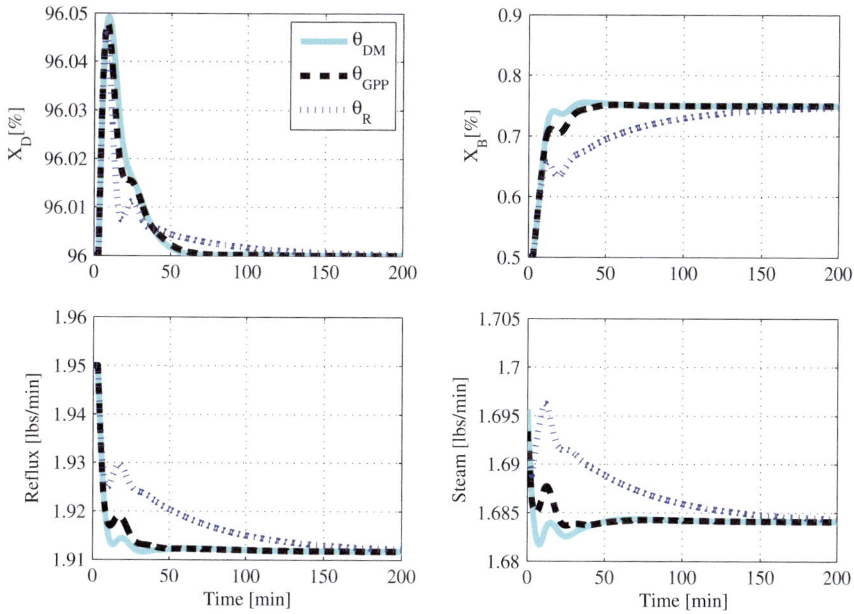

Fig. 5.7 Time response comparison for a change in the bottom product setpoint X_B from 0.5 to 0.75 % (Test 2)

Table 5.7 Performance X_D for a change in the bottom product setpoint X_B from 0.5 to 0.75 % (Test 2)

	IAE	ITAE	ISE	ITSE	Overshoot	$t_{98\,\%}$
$\boldsymbol{\theta}_{DM}$	8.91	171.23	0.27	3.46	0.05	54.31
$\boldsymbol{\theta}_{GPP}$	8.18	161.68	0.22	2.71	0.05	59.24
$\boldsymbol{\theta}_{R}$	8.33	308.02	0.15	2.02	0.05	131.46

Table 5.8 Performance X_B for a change in the bottom product setpoint X_B from 0.5 % to 0.75 % (Test 2)

	IAE	ITAE	ISE	ITSE	Overshoot	$t_{98\,\%}$
$\boldsymbol{\theta}_{DM}$	23.54	201.62	4.10	15.30	0.70	39.29
$\boldsymbol{\theta}_{GPP}$	28.05	290.49	4.29	19.54	0.20	38.24
$\boldsymbol{\theta}_{R}$	80.09	3.53 + 3	7.87	149.95	0.00	172.70

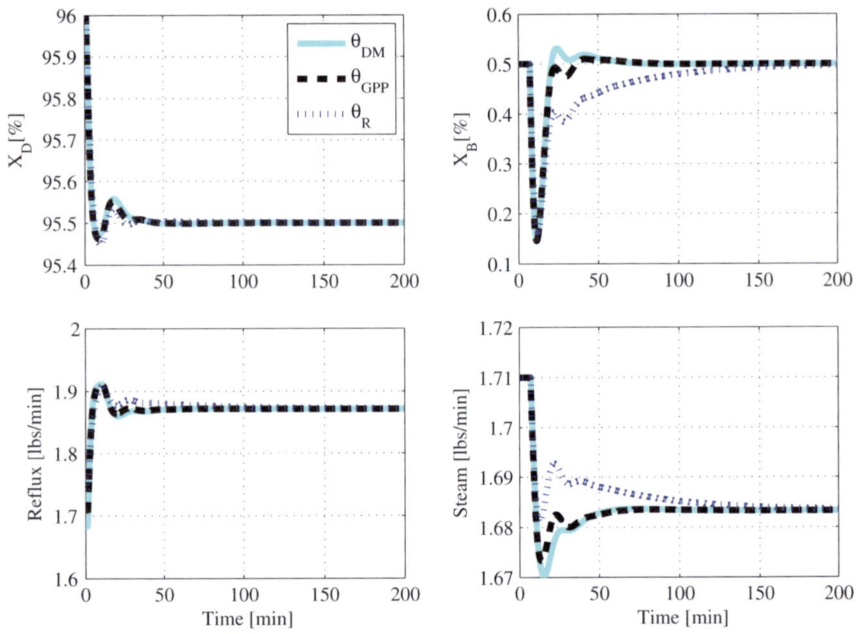

Fig. 5.8 Time response comparison for a change in the overhead product setpoint X_D from 96.0 to 95.5 % (Test 3)

Table 5.9 Performance X_D for a change in the overhead product setpoint X_D from 96.0 to 95.5 % (Test 3)

	IAE	ITAE	ISE	ITSE	Overshoot	$t_{98\,\%}$
θ_{DM}	22.05	229.27	5.30	11.26	0.52	31.73
θ_{GPP}	21.40	189.35	5.38	9.91	0.52	26.64
θ_R	23.03	326.35	5.86	9.93	0.52	22.86

Table 5.10 Performance X_B for a change in the overhead product setpoint X_D from 96.0 to 95.5 % (Test 3)

	IAE	ITAE	ISE	ITSE	Overshoot	$t_{98\,\%}$
θ_{DM}	33.61	602.80	7.44	91.88	6.23	56.29
θ_{GPP}	33.33	594.83	7.49	93.06	1.96	59.02
θ_R	82.48	$3.7e+3$	10.84	234.28	0.00	158.58

5.5 Conclusions

In this chapter, a multivariable controller was tuned by means of a MOOD procedure. For this problem, it was necessary to state design objectives for each control loop, leading to a many-objectives optimization instance. To overcome such an issue, a compact and pertinent Pareto Front approximation was calculated, in order to select a preferable solution.

In this case, a *reference tuning rule* controller was used, in order to provide additional meaning to indicators as IAE and TV. Such a controller might be a previously tuned controller, a well-known tuning rule or the expected performance of using another controller. This allows one to state a preference matrix according with the improvements of such reference controller.

Also, structural difference between three different approaches to deal with the MOP were shown. Preferring one over another will rely on the confidence and desires of the DM:

- If the DM is confident with the preference matrix, and needs a solutions right now, a single-objective approach could be useful, using the GPP index.
- If the DM is seeking to improve its knowledge regarding trade-off among conflicting objectives, then approximating a dense Pareto Front will be useful.
- If the DM is seeking to implement a desirable solution, but would like to analyze the trade-off in the surroundings of the preferable region, then a compact and pertinent Pareto Front will be useful.

References

1. Berry MW (1973) Terminal composition control of a binary distillation column. Master's thesis, Department of Chemical and Petroleum Engineering, University of Alberta., Edmonton, Alberta
2. Das S, Suganthan PN (2010) Differential evolution: a survey of the state-of-the-art. IEEE Trans Evol Comput 99:1–28
3. dos Santos Coelho L, Pessôa MW (2011) A tuning strategy for multivariable PI and PID controllers using differential evolution combined with chaotic zaslavskii map. Expert Syst Appl 38(11):13694–13701
4. Iruthayarajan MW, Baskar S (2009) Evolutionary algorithms based design of multivariable PID controller. Expert Syst Appl 36(5):9159–9167
5. Iruthayarajan MW, Baskar S (2010) Covariance matrix adaptation evolution strategy based design of centralized PID controller. Expert Syst Appl 37(8):5775–5781
6. Luyben WL (1986) Simple method for tuning SISO controllers in multivariable systems. Indus Eng Chem Process Des 25:654–660
7. Menhas MI, Wang L, Fei M, Pan H (2012) Comparative performance analysis of various binary coded PSO algorithms in multivariable PID controller design. Expert Syst Appl 39(4):4390–4401
8. Reynoso-Meza G, Sanchis J, Blasco X, García-Nieto S (2014) Physical programming for preference driven evolutionary multi-objective optimization. Appl Soft Comput 24:341–362
9. Reynoso-Meza G, Sanchis J, Blasco X, Herrero JM (2012) Multiobjective evolutionary algortihms for multivariable PI controller tuning. Expert Syst Appl 39:7895–7907

10. Reynoso-Meza G, Sanchis J, Blasco X, Martínez M (2010) Design of continuous controllers using a multiobjective differential evolution algorithm with spherical pruning. In: Applications of evolutionary computation. Springer, pp 532–541
11. Storn R (2008) SCI: Differential evolution research: trends and open questions, vol LNCS 143. Springer, Heidelberg, pp 1–31
12. Storn R, Price K (1997) Differential evolution: a simple and efficient heuristic for global optimization over continuous spaces. J Global Optim 11:341–359
13. Wood RK, Berry MW (1973) Terminal composition control of a binary distillation column. Chem Eng Sci 28(9):1707–1717

Chapter 6
Comparing Control Structures from a Multiobjective Perspective

Abstract This chapter will illustrate the tools presented in previous chapters for the analysis and comparison of different design concepts. In particular, three different control structures (PI, PID and GPC) will be compared, analysing their benefits and drawbacks within a multiobjective approach. First, a two objective approach, where robustness and disturbance rejection are analyzed, will be developed. Later, a third objective will be added related to setpoint tracking. Since PI design concept has only two parameters to be tuned, the PID design concept will be set with a derivative gain K_d depending on other controller parameters for a fair comparison. Regarding the Generalized Predictive Controller (GPC) all parameters except prediction horizon and filter parameter will be fixed. Development of the example let the reader know how the tools can help to compare different control structures and how to choose the parameters for the best controller from the point of view of DM within a MOOD approach.

6.1 Introduction

It is common in control engineering to have a variety of control structures to control a process, without having a clear idea about the best choice. In an MO context, the good selection has some degree of subjectivity depending on the engineer's preferences. Tools for solutions comparison are always welcome to give additional information that leads to a more reliable selection. Different design alternatives or design concepts will have an associated Pareto Set, corresponding to the non-dominated design parameter values for each structure.

In this chapter, selection of an adequate control structure among three alternatives (PI, PID, and a Generalized Predictive Controller GPC) and their corresponding particular values will be illustrated.

G. Reynoso Meza et al., *Controller Tuning with Evolutionary Multiobjective Optimization*, Intelligent Systems, Control and Automation: Science and Engineering 85, DOI 10.1007/978-3-319-41301-3_6

6.2 Model and Controllers Description

In [1] is presented the idea that derivative action is useful for first order plus time delay processes. For the extreme case of integrator plus time delay model:

$$P(s) = \frac{e^{-s}}{s}, \tag{6.1}$$

three digital controllers, with control period of $T_s = 0.5$, will be designed and compared:

1. A digital 1-DOF PI controller over error signal (Fig. 6.1), with bilinear approximation to its derivative:

$$u(t) = C_{PI}(z^{-1})e(t) = \left(K_c + K_i \frac{T_s}{2} \frac{1+z^{-1}}{1-z^{-1}}\right) e(t). \tag{6.2}$$

2. A digital 2-DOF PID controller (Fig. 6.2) with derivative filter, derivative effect only on the output and bilinear approximation to its derivative:

$$\begin{aligned} u(t) &= C_{PID_1}(z^{-1})e(t) + C_{PID_2}(z^{-1})y(t) \\ &= \left(K_c + K_i \frac{T_s}{2} \frac{1+z^{-1}}{1-z^{-1}}\right) e(t) + \left(\frac{K_d}{\frac{K_d}{N} + \frac{T_s}{2}\frac{1+z^{-1}}{1-z^{-1}}}\right) y(t). \end{aligned} \tag{6.3}$$

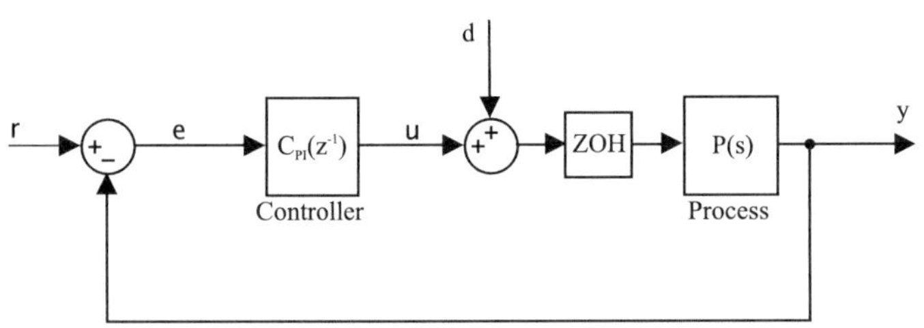

Fig. 6.1 Control loop for a 1-DOF digital PI control

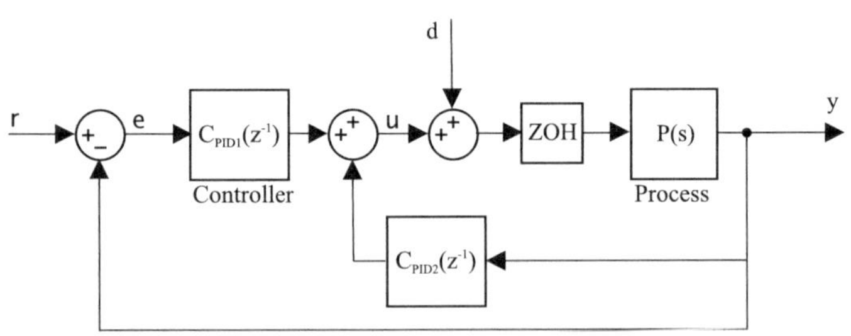

Fig. 6.2 Control loop for a 2-DOF digital PID control

If K_c, K_i and K_d were tuned freely, performances would be improved with respect to previous PI controller. In this case, the designer chooses a tuning rule that relates integral and derivative parts[1] so there are only two parameters to tune (same as in 1-DOF PI). Therefore, K_c and K_i have to be designed and K_d is set as $K_d = K_c^2/4/K_i$. The derivative filter parameter is fixed to $N = 2$.

3. A Generalized Predictive Controller (GPC). Its formulation with quadratic cost index has been extensively developed in [2–4]. GPC uses the known CARIMA time series model, where stochastic components are included into the system description as random effects:

$$y(t) = \frac{B(z^{-1})}{A(z^{-1})}u(t-1) + \frac{T(z^{-1})}{\Delta A(z^{-1})}\xi(t) \tag{6.4}$$

where $u(t)$ and $y(t)$ are the process input and output respectively, $B(z^{-1})$ and $A(z^{-1})$ are the numerator and denominator polynomials of the discrete transfer function of the process, $\xi(t)$ is assumed white noise, Δ operator is added to avoid steady state error and $T(z^{-1})$ polynomial is used to filter disturbances and model uncertainties (in fact, $T(z^{-1})$ could be considered as part of the controller rather than a part of the model and can be tuned in different ways for that purpose).
The GPC control law is calculated through the optimization of the following cost index:

$$J(\Delta u) = \sum_{i=N_1}^{N_2} \alpha[y(t+i) - r(t)]^2 + \sum_{j=1}^{N_u} \lambda[\Delta u(t+j-1)]^2 \tag{6.5}$$

where $N = N_2 - N_1 + 1$ is the prediction horizon, N_u is the control horizon, α is the prediction error weighting factor, λ is the control weighting factor, $r(t)$ is the setpoint, and $[\ \Delta u(t)\ \ \Delta u(t+1)\ \cdots\ \Delta u(t+N_u-1)\ ]^T$ are the future control movements. Different GPC tuning approaches has been developed [3, 5] where appropriate values for these parameters can be found. However, since a GPC has more parameters to tune than a PID structure: N_1, N_2, N_u, α, λ and polynomial $T(z^{-1})$, they will be limited to two as in the PI and PID controllers.
Optimizing index (6.5) gives a linear control law which can be posed as a linear controller[2] (Fig. 6.3):

$$u(t) = \frac{T(z^{-1})\left(H_0 r(t) - \frac{S(z^{-1})}{T(z^{-1})}y(t)\right)}{\left(T(z^{-1}) + R(z^{-1})z^{-1}\right)\Delta}. \tag{6.6}$$

[1] As in Ziegler and Nichols tuning rules, $T_i = 4T_d$ for the ISA PID [6]. For parallel form this relation is transformed in $K_d = K_c^2/4/K_i$.

[2] Using the predictive control principle of *Receding Horizon* where several control movements are calculated each time period but only the first one, $\Delta u(t)$, is applied to the process.

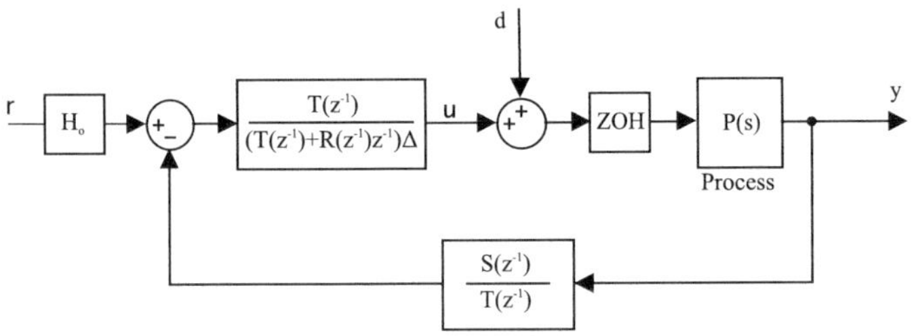

Fig. 6.3 Control structure for GPC

GPC will use process model $P(s)$ expressed as CARIMA model, so

$$y(t) = \frac{0.5z^{-1}}{1 - z^{-1}} z^{-2} u(t-1) + \frac{1 - \alpha_f z^{-1}}{(1 - z^{-1})^2} \xi(t). \tag{6.7}$$

Notice that polynomial $T(z^{-1})$ is adjusted as a first order filter, $1 - \alpha_f z^{-1}$, where α_f will be a tuning parameter. The remaining parameters are set to: $\alpha = 1, \lambda = 0$, $N_u = 1$, $N_1 = 2 + 1$ (since the discrete process presents a delay of two samples), except N_2 which also will be tuned.

For unstable processes (our case) high values of λ could produce unstable closed loop behaviour. On the other hand λ avoids aggressive control actions and overshoot, but that effect can be obtained too modifying N_2 adequately. So that λ is set to zero and N_2 tuned. Selecting N_u equal to the order of polynomial $A(z^{-1})$ is a good compromise for robustness and performance, so that $N_u = 1$.

6.3 The MOOD Approach

In this section, two comparison scenarios will be proposed. First, a 2D MOP will be stated and the three control structures will be compared (and tuned) discussing the benefits and limitations of each design concept. Afterwards, the problem will be extended with a third objective where Level Diagrams will play an important role showing main characteristics of each design concept and helping in the analysis of the Pareto solution.

6.3.1 Two Objectives Approach

Performance related to disturbance rejection is considered by means of the IAE_d index ($d(t)$ is a unitary step at $t = 0$). Besides, robustness is taken into account by means of the maximum of sensitivity function M_s.

The following three concepts (PI, PID and GPC) will be compared:

1. MO problem for PI tuning. Controller represented in Eq. (6.2).

$$\boldsymbol{\Theta}_{PI} = arg\left(\min_{\theta_{PI}} \boldsymbol{J}(\boldsymbol{\theta}_{PI})\right) = arg\left(\min_{\theta_{PI}}[M_s(\boldsymbol{\theta}_{PI}), IAE_d(\boldsymbol{\theta}_{PI})]\right) \tag{6.8}$$

with

$$\boldsymbol{\theta}_{PI} = [K_c, K_i] \tag{6.9}$$

subject to[3]:

$$\begin{aligned} 0 < K_c &\leq 1 \\ 0 \leq K_i &\leq 1 \\ 1 \leq M_s &\leq 2 \\ t_{98\,\%} \leq t_{sim} &= 100 \text{ s.} \end{aligned} \tag{6.10}$$

2. MO problem for PID tuning. Controller represented in Eq. (6.3).

$$\boldsymbol{\Theta}_{PID} = arg\left(\min_{\theta_{PID}} \boldsymbol{J}(\boldsymbol{\theta}_{PID})\right) = arg\left(\min_{\theta_{PID}}[M_s(\boldsymbol{\theta}_{PID}), IAE_d(\boldsymbol{\theta}_{PID})]\right) \tag{6.11}$$

with

$$\boldsymbol{\theta}_{PID} = [K_c, K_i] \tag{6.12}$$

subject to

$$\begin{aligned} 0 < K_c &\leq 1 \\ 0 \leq K_i &\leq 1 \\ K_d &= \frac{K_c^2}{4 \cdot K_i} \\ 1 \leq M_s &\leq 2 \\ t_{98\,\%} \leq t_{sim} &= 100 \text{ s.} \end{aligned} \tag{6.13}$$

3. MO problem for GPC tuning. Controller represented in Eq. (6.6).

$$\boldsymbol{\Theta}_{GPC} = arg\left(\min_{\theta_{GPC}} \boldsymbol{J}(\boldsymbol{\theta}_{GPC})\right) = arg\left(\min_{\theta_{GPC}}[M_s(\boldsymbol{\theta}_{GPC}), IAE_d(\boldsymbol{\theta}_{GPC})]\right) \tag{6.14}$$

with

$$\boldsymbol{\theta}_{GPC} = [N_2, \alpha_f] \tag{6.15}$$

[3]The last two constraints have been added to increase pertinency of the solutions since outside these limits they are not interesting at all. t_{sim} is the closed loop simulation time over which the objectives are calculated.

Table 6.1 Parameters used for sp-MODE. Further details in Chap. 3

Parameter	Value
Evolutionary mechanism	
F (scaling factor)	0.5
Cr (crossover rate)	0.9
N_p (population)	50
Pruning mechanism	
β^{ϵ} (Arcs)	40

subject to

$$\begin{aligned} 5 \le N_2 &\le 50 \\ 0 \le \alpha_f &\le 0.99 \\ 1 \le M_s &\le 2 \\ t_{98\,\%} \le t_{sim} &= 100 \text{ s.} \end{aligned} \tag{6.16}$$

The sp-MODE algorithm is used to solve the MOPs stated above, using parameters of Table 6.1. Figure 6.4 shows the Pareto Fronts and Pareto Sets of the three MO

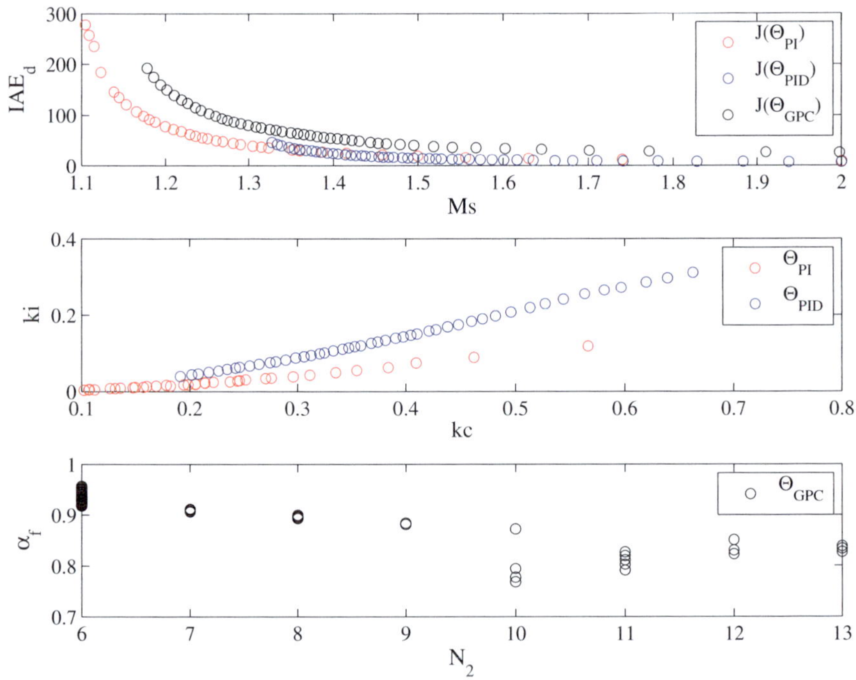

Fig. 6.4 Pareto Fronts $\boldsymbol{J}(\boldsymbol{\Theta}_{PI})$, $\boldsymbol{J}(\boldsymbol{\Theta}_{PID})$, $\boldsymbol{J}(\boldsymbol{\Theta}_{GPC})$ and Pareto Sets $\boldsymbol{\Theta}_{PI}$, $\boldsymbol{\Theta}_{PID}$, $\boldsymbol{\Theta}_{GPC}$ approximations for the three MOPs

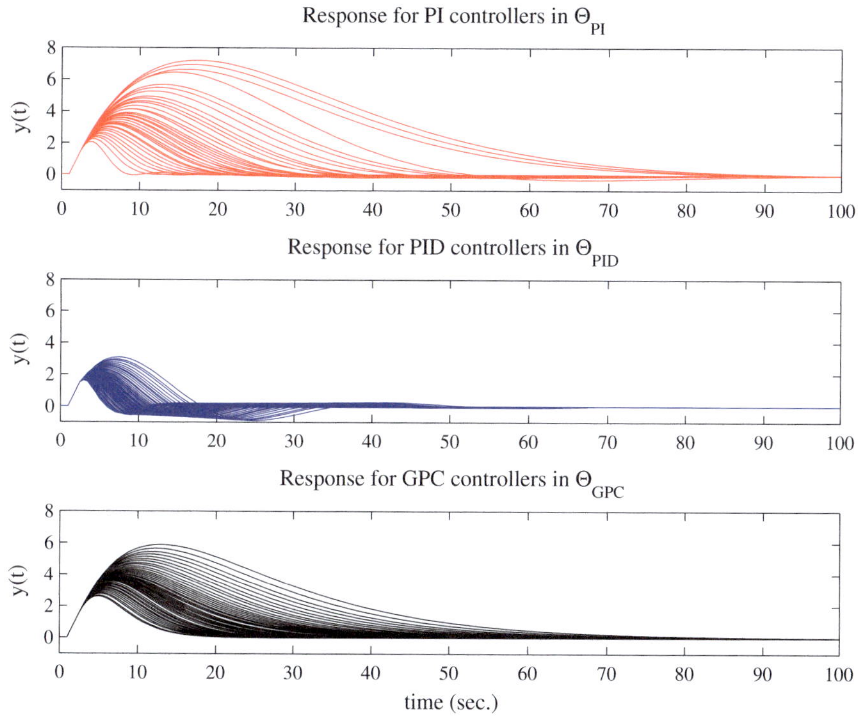

Fig. 6.5 Output response $y(t)$ for an unitary step in $d(t)$ at $t = 0$ for controllers in $\boldsymbol{\Theta}_{PI}$, $\boldsymbol{\Theta}_{PID}$, $\boldsymbol{\Theta}_{GPC}$

optimizations. Figure 6.5 shows the closed loop responses $y(t)$ applying the solutions in each Pareto Set.

Comparing $\boldsymbol{J}(\boldsymbol{\Theta}_{PI})$, $\boldsymbol{J}(\boldsymbol{\Theta}_{PID})$ and $\boldsymbol{J}(\boldsymbol{\Theta}_{GPC})$ you can realize that PI and PID controllers dominate GPC ones. The minimum value of IAE_d for a GPC controller is 25.18 with $M_s = 2$, whilst PI or PID controllers with same IAE_d reach $M_s \simeq 1.4$ (more robust). Notice that a GPC controller is conceived using a CARIMA model where the disturbance is a filtered white noise $\xi(t)$, whilst an IAE_d index is calculated when a unitary step is applied in disturbance $d(t)$.

On the other hand, neither PI controllers completely dominate PIDs in all Pareto Front, or vice versa. PID controllers dominate (slightly) PIs when $M_s > 1.4$ whilst PI controllers dominate PIDs when $M_s < 1.4$. Table 6.2 compares performance in IAE_d for different values of M_s.

Our main conclusion after this analysis is that the GPC controller is the worst design concept while the PID controller has slightly better performance than PI one, although the PI simplicity against PID one makes the DM choosing the PI controller as the preferred concept. Finally, a PI controller with $K_c = 0.354$ and $T_i = 0.055$ resulting in $M_s = 1.5$ and $IAE_d = 18.7$ is selected.

Table 6.2 PI, PID and GPC controllers comparative performance

M_s	IAE_d	Controller
2	6.6	PID
2	8.62	PI
2	25.18	GPC
1.5	14.8	PID
1.5	18.7	PI
1.5	40	GPC
1.32	35.5	PI
1.32	45.88	PID
1.32	69.7	GPC

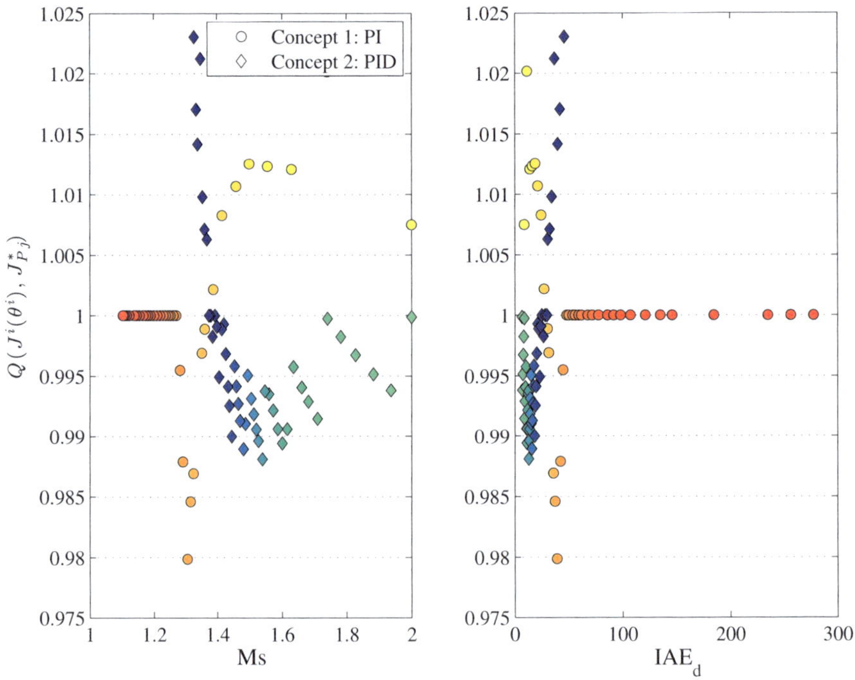

Fig. 6.6 Comparison of PI and PID concepts using LD and Q indicator

In MOPs with two objectives, plots like Fig. 6.4 are useful for comparing different design concepts. Let's see if LD with quality indicator Q supplies the same type of information. In Fig. 6.6 a comparison of PI and PID concepts are depicted. Notice how values of $Q = 1$ (for $M_s \in [1.1 \ldots 1.3]$ and $IAE_d \in [50 \ldots 300]$) indicates that PID is not covering this part of the objective space. For $M_s \in [1.3 \ldots 1.4]$, values of $Q < 1$ for PI concept and $Q > 1$ for PID indicating that PI dominates

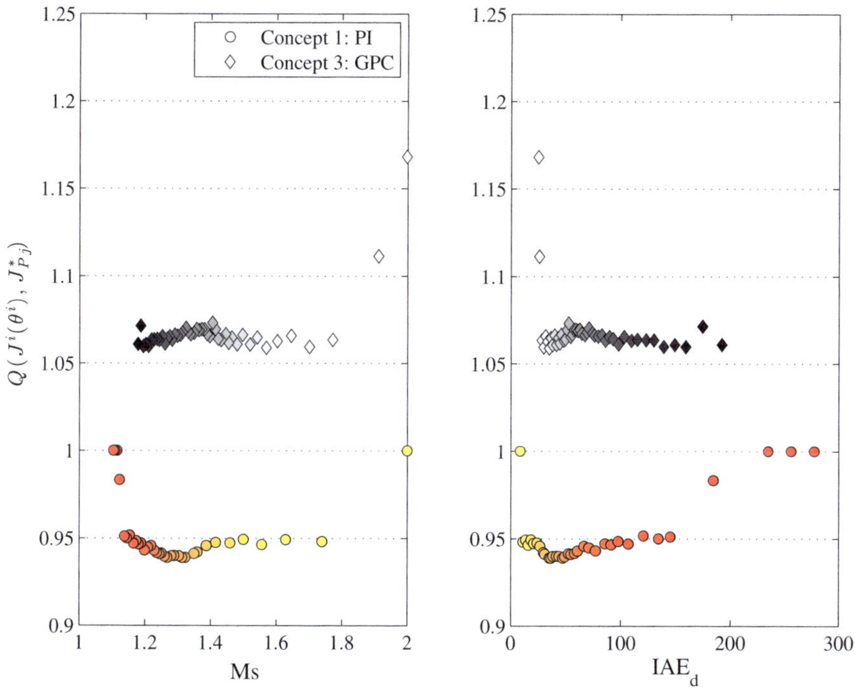

Fig. 6.7 Comparison of PI and GPC concepts using level diagrams and Q indicator

PID in this area. On the other hand for $M_s \in [1.4\ldots2]$, indicator $Q > 1$ for PI concept and $Q < 1$ for PID one indicating PID dominates PI in this area. Remarks that the dominance is not high, the values of the Q indicator are close to 1. Similar conclusions can be obtained analysing an IAE_d Level Diagram.

In a similar way, Fig. 6.7 compares PI and GPC concepts, where values of $Q < 1$ for PI concept and $Q > 1$ for GPC indicates that PI dominates completely GPC. For PID and GPC concepts, Fig. 6.8 depicts values of $Q < 1$ for PID and $Q > 1$ for GPC showing that PID concept dominates GPC one when $M_s < 1.35$ and $IAE_d < 75$ values. However, for $M_s > 1.35$ and $IAE_d < 75$, indicator $Q = 1$ for GPC concept meaning that these areas are not reached by the PID concept.

6.3.2 Three Objectives Approach

Beside previous IAE_d and M_s a third objective is added related to the set-point response by means of IAE_r (when $r(t)$ is a unitary step at $t = 0$). Now a different MOP is considered and again the following problems are solved:

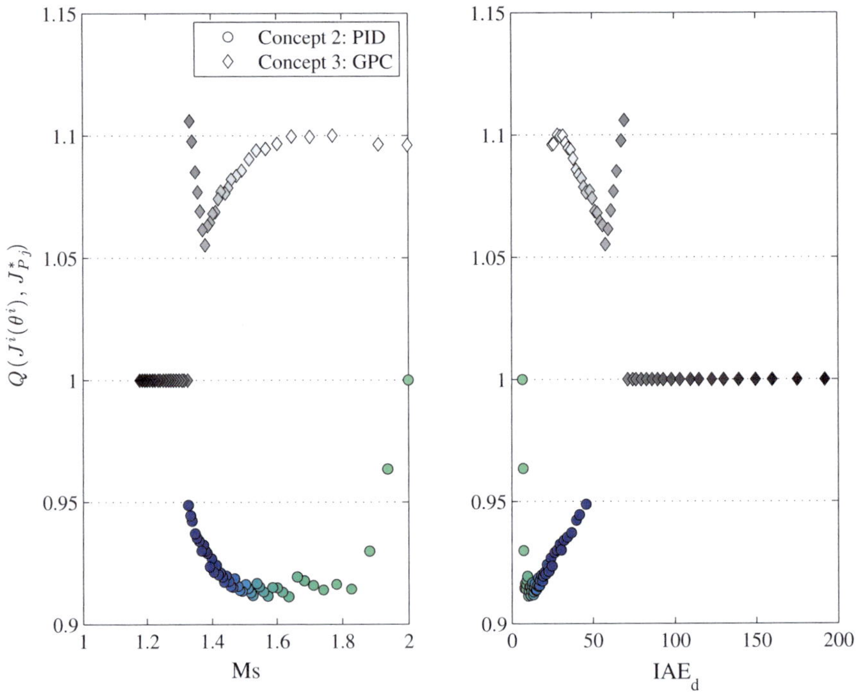

Fig. 6.8 Comparison of PID and GPC concepts using level diagrams and Q indicator

1. MO problem for PI tuning. Controller represented in Eq. (6.2).

$$\boldsymbol{\Theta}_{PI} = arg\left(\min_{\theta_{PI}} \boldsymbol{J}(\boldsymbol{\theta}_{PI})\right) = arg\left(\min_{\theta_{PI}}[M_s(\boldsymbol{\theta}_{PI}), IAE_d(\boldsymbol{\theta}_{PI}), IAE_r(\boldsymbol{\theta}_{PI})]\right) \tag{6.17}$$

with

$$\boldsymbol{\theta}_{PI} = [K_c, K_i] \tag{6.18}$$

subject to:

$$\begin{aligned} 0 < K_c &\leq 1 \\ 0 \leq K_i &\leq 1 \\ 1 \leq M_s &\leq 2 \\ t_{98\,\%} \leq t_{sim} &= 100\text{s}. \end{aligned} \tag{6.19}$$

2. MO problem for PID tuning. Controller represented in Eq. (6.3).

$$\boldsymbol{\Theta}_{PID} = \min_{\theta_{PID}} \boldsymbol{J}(\boldsymbol{\theta}_{PID}) = \min_{\theta_{PID}} [M_s(\boldsymbol{\theta}_{PID}), IAE_d(\boldsymbol{\theta}_{PID}), IAE_r(\boldsymbol{\theta}_{PID})] \quad (6.20)$$

with

$$\boldsymbol{\theta}_{PID} = [K_c, K_i] \quad (6.21)$$

subject to:

$$\begin{aligned} 0 < K_c &\leq 1 \\ 0 \leq K_i &\leq 1 \\ K_d &= \frac{K_c^2}{4 \cdot K_i} \\ 1 \leq M_s &\leq 2 \\ t_{98\,\%} \leq t_{sim} &= 100\,\text{s} \end{aligned} \quad (6.22)$$

3. MO problem for GPC tuning. Controller represented in Eq. (6.6).

$$\boldsymbol{\Theta}_{GPC} = \min_{\theta_{GPC}} \boldsymbol{J}(\boldsymbol{\theta}_{GPC}) = \min_{\theta_{GPC}} [M_s(\boldsymbol{\theta}_{GPC}), IAE_d(\boldsymbol{\theta}_{GPC}), IAE_r(\boldsymbol{\theta}_{GPC})] \quad (6.23)$$

with

$$\boldsymbol{\theta}_{GPC} = [N_2, \alpha_f] \quad (6.24)$$

subject to:

$$\begin{aligned} 5 \leq N_2 &\leq 50 \\ 0 \leq \alpha_f &\leq 0.99 \\ 1 \leq M_s &\leq 2 \\ t_{98\,\%} \leq t_{sim} &= 100\,\text{s}. \end{aligned} \quad (6.25)$$

sp-MODE algorithm is parameterized as in the previous case and used to solve the three MOPs. Figure 6.9 shows the Pareto Fronts and Pareto Sets resulting from the three optimizations (now is more difficult to analyze and compare Pareto Fronts $\boldsymbol{J}(\boldsymbol{\Theta}_{PI})$, $\boldsymbol{J}(\boldsymbol{\Theta}_{PID})$ and $\boldsymbol{J}(\boldsymbol{\Theta}_{GPC})$ in a 3D space.) whilst Fig. 6.10 shows the output response $y(t)$ for each controller belonging to these sets when set point $r(t)$ and disturbance $d(t)$ step changes are applied.

Regarding Fig. 6.10 notice that GPC controllers have very good response to set point changes but not to disturbances ones.

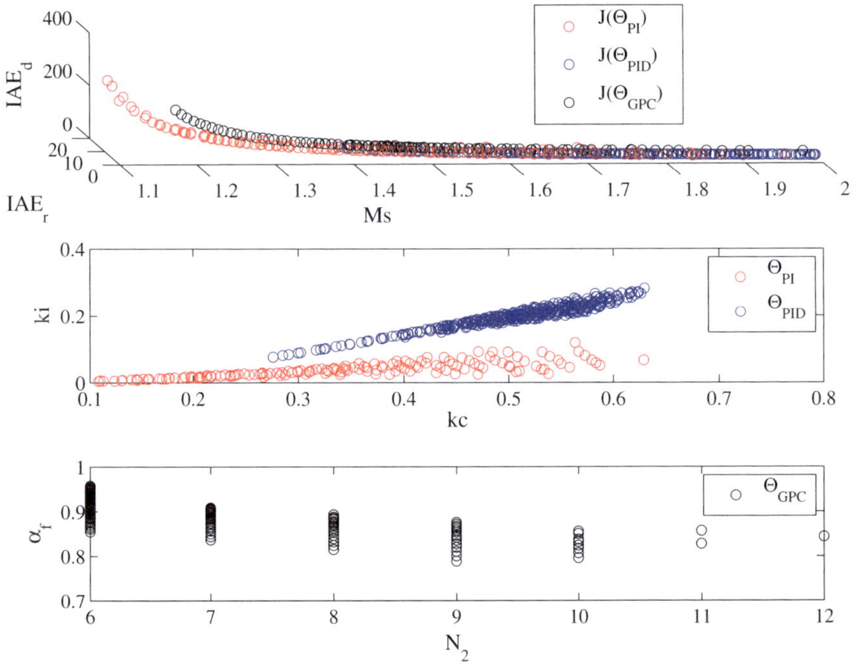

Fig. 6.9 Pareto Fronts $\boldsymbol{J}(\boldsymbol{\Theta}_{PI})$, $\boldsymbol{J}(\boldsymbol{\Theta}_{PID})$, $\boldsymbol{J}(\boldsymbol{\Theta}_{GPC})$ and Pareto Sets $\boldsymbol{\Theta}_{PI}$, $\boldsymbol{\Theta}_{PID}$, $\boldsymbol{\Theta}_{GPC}$ for the three objective problems presented

Again, let's see how LD allows us a deep analysis of the Pareto Fronts. Figure 6.11 shows the LD for PI design concept with ∞-norm. LD has been colored in such a way that the darker the color point the lower is the value of M_s (same color for all diagrams). Notice that M_s objective is in opposition to IAE_r and IAD_d and that controllers with good performance in IAE_r and IAD_d are those with $K_c \in [0.4 \cdots 0.6]$ and $K_i \in [0.04 \cdots 0.1]$. Several options are available: a controller with the minimum $\|J\|_\infty$, (PI_1) gives $\|J\|_\infty = 0.23$ so that the loss of performance (in any of the three objectives) does not exceed 23 % with respect to the complete range of values of the approximated Pareto Front. Other selections can be PI_2 with $M_s = 1.5$ and PI_3 with $M_s = 2$ (see Table 6.3).

Regarding the PID design concept, Fig. 6.12 shows the Pareto Front $\boldsymbol{J}(\boldsymbol{\Theta}_{PID})$ represented with LD. Again, M_s index is in opposition to IAE_r and IAD_d and controllers with good performances in IAE_r and IAD_d take their parameters from $K_c \in [0.5 \cdots 0.6]$, $K_i \in [0.2 \cdots 0.28]$ and $K_d \in [0.32 \cdots 0.38]$. A PID controller with minimum $\|J\|_\infty$ value is selected (PID_1) as well as PID_2 with $M_s = 1.5$ and PID_3 with $M_s = 2$ (see Table 6.4).

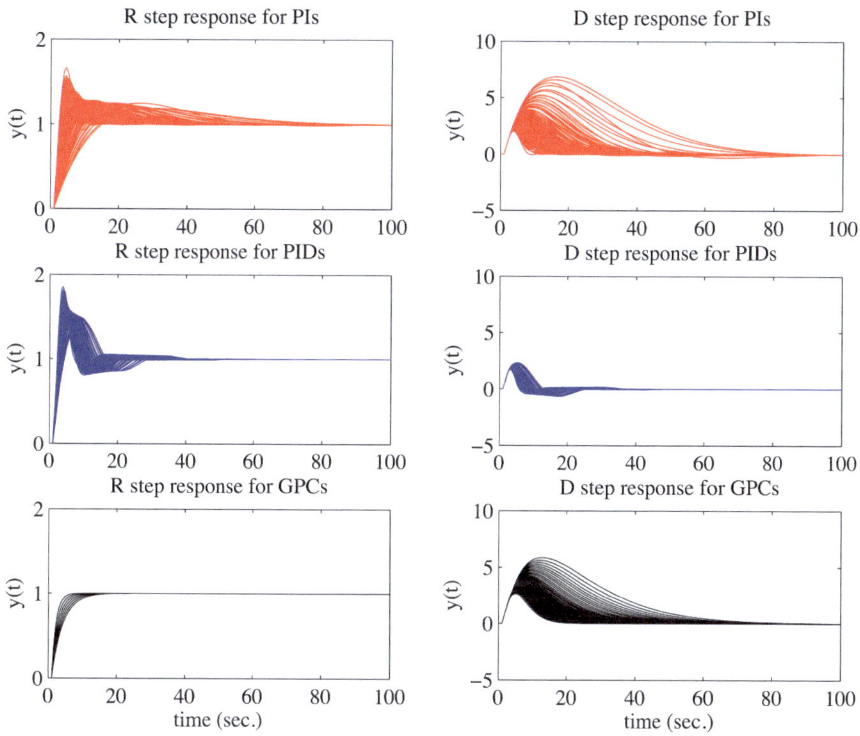

Fig. 6.10 Closed loop response for each controller in $\boldsymbol{\Theta}_{PI}$, $\boldsymbol{\Theta}_{PID}$, $\boldsymbol{\Theta}_{GPC}$. Output $y(t)$ for an unitary step in $r(t)$ at $t = 0$ (*left*). Output $y(t)$ for an unitary step in $d(t)$ at $t = 0$ (*right*)

Finally, Fig. 6.13 shows the LD for the GPC design concept. In this case, objective M_s is in opposition to IAD_d one but not with IAD_r. Now IAD_d and IAD_r are in opposition, so there is not a GPC controller with good performance in disturbance rejection and sep-point response simultaneously. Controllers that produce good M_s values have $N_2 \in [6 \cdots 8]$ and $\alpha_f \in [0.88 \cdots 0.96]$. Finally, selection of GPC with lowest $\|J\|_\infty$ (GPC_1), GPC_2 with $M_s = 1.5$ and GPC_3 with $M_s = 2$ are shown in Table 6.5.

In order to compare PI, PID and GPC concepts using LD, the $\boldsymbol{J}(\boldsymbol{\Theta}_{PI})$, $\boldsymbol{J}(\boldsymbol{\Theta}_{PID})$ and $\boldsymbol{J}(\boldsymbol{\Theta}_{GPC})$ Pareto Fronts have been joined and represented in the same figure using $\|J\|_\infty$ and $\|J\|_2$ norms separately (Fig. 6.14). A PID controller can not achieve values lower than 1.4 in M_s (unless $t_{98\,\%}$ constraint would be unsatisfied) whilst PI and GPC can get values in M_s lower than 1.2. The best performance in IAE_r is obtained with GPC followed by PI. On the other hand, the best IAE_d performance is obtained with PID controllers followed by PI ones (with similar performance) and finally with GPC ones.

Selected solutions PI_1, PID_1 and GPC_1 are compared in Fig. 6.15. As well as PID_1 controller obtains a good disturbance rejection (maximum set-point deviation

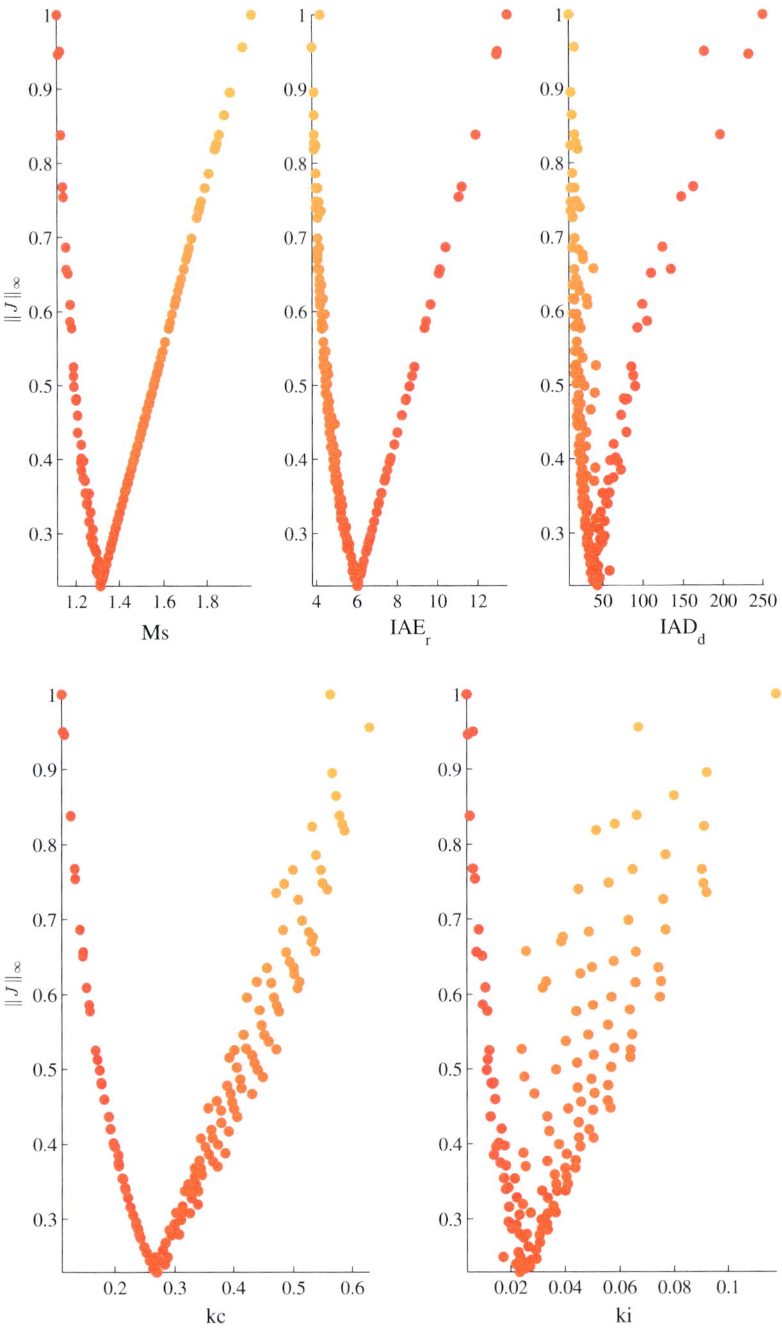

Fig. 6.11 Level Diagrams for $\boldsymbol{J}(\boldsymbol{\Theta}_{PI})$ and $\boldsymbol{\Theta}_{PI}$. $\|J\|_\infty$ is used for y-axis synchronization

Table 6.3 Comparison between different Pareto solutions for PI controller

PI	K_c	K_i	M_s	IAE_r	IAE_d
PI_1	0.27	0.023	1.31	6	42.9
PI_2	0.40	0.033	1.5	4.65	29.96
PI_3	0.56	0.011	2	4.18	8.64

Table 6.4 Comparison between different Pareto solutions for PID controller

PID	K_c	K_i	K_d	M_s	IAE_r	IAE_d
PID_1	0.44	0.16	0.30	1.61	5.93	11.51
PID_2	0.38	0.13	0.27	1.5	6.58	14.37
PID_3	0.61	0.25	0.37	2	4.86	6.84

Table 6.5 Comparison between different Pareto solutions for GPC controller

GPC	N_2	α_f	M_s	IAE_r	IAE_d
GPC_1	6	0.91	1.36	3.21	63.05
GPC_2	10	0.85	1.5	4.59	40.23
GPC_3	9	0.78	2	4.25	25.5

is lower than 2), its set-point response shows excessive overshoot and t_e (bigger than 50 % and 25 s respectively). Just the opposite happens with GPC_1 with a very desirable set-point response (no overshoot at all and $t_{98\,\%} \simeq 6$ s) but with poor disturbance rejection (maximum output deviation is near 4). PI_1 presents an intermediate performance for set-point response (overshoot $\simeq 20$ % and $t_{98\,\%}$ $\simeq 30$ s) for and a maximum output deviation of $\simeq 3$ when disturbance appears.

Similar conclusions are obtained when particular solutions with $M_s = 1.5$ (PI_2, PID_2 and GPC_2) are compared. Regarding solutions with $M_s = 1$, PI_3 and PID_3 have similar performances but GPC_3 gets a good set-point response (no overshoot and $t_e \simeq 10$ s with worse disturbance rejection than the others but improved respect to GPC_1 and GPC_2 (maximum output deviation is lower than 3 and disturbance is rejected before 20 s). Therefore GPC_3 could be selected if the DM considers that set-point response is more relevant than disturbance rejection whilst PI_2 could be selected as a good compromise controller.

Let's use LD together with quality indicator Q to compare the alternative control structures used in this problem (Figs. 6.16, 6.17 and 6.18). Notice how in PI vs PID (Fig. 6.16) Q indicator is equal to 1, so that no concept dominates the other. The two concepts cover different parts of the objective space. Same conclusions are obtained (see Fig. 6.18) when PID and GPC concepts are compared.

A different situation appears when PI vs GPC comparison (Fig. 6.17). Notice that for low values of IAE_d (left sub-plot), PI concept clearly dominates GPC ($Q < 1$ for PI controllers and $Q > 1$ for GPC ones) and that GPC dominates PI for low values of IAE_r (center sub-plot).

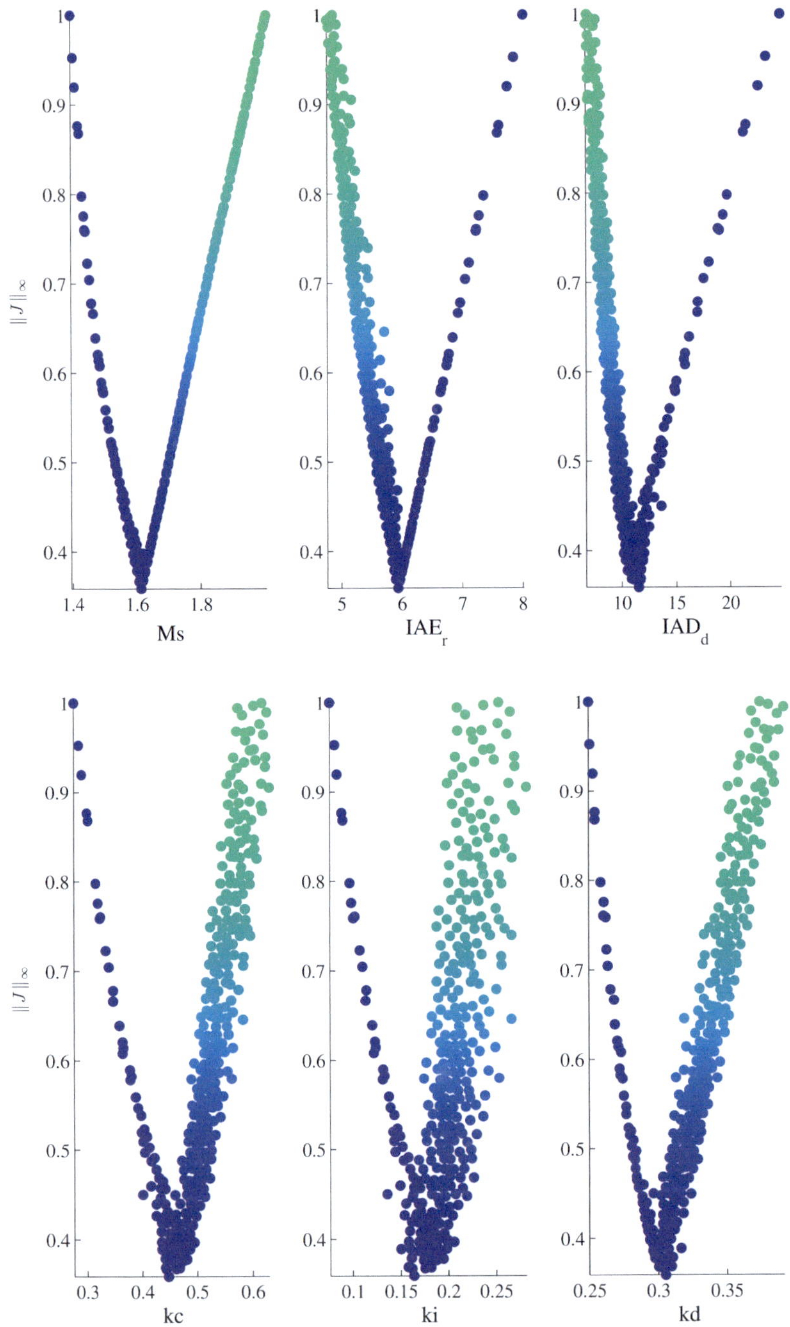

Fig. 6.12 Level diagrams for $\boldsymbol{J}(\boldsymbol{\Theta}_{PID})$ and $\boldsymbol{\Theta}_{PID}$. $\|J\|_\infty$ is used for y-axis synchronization

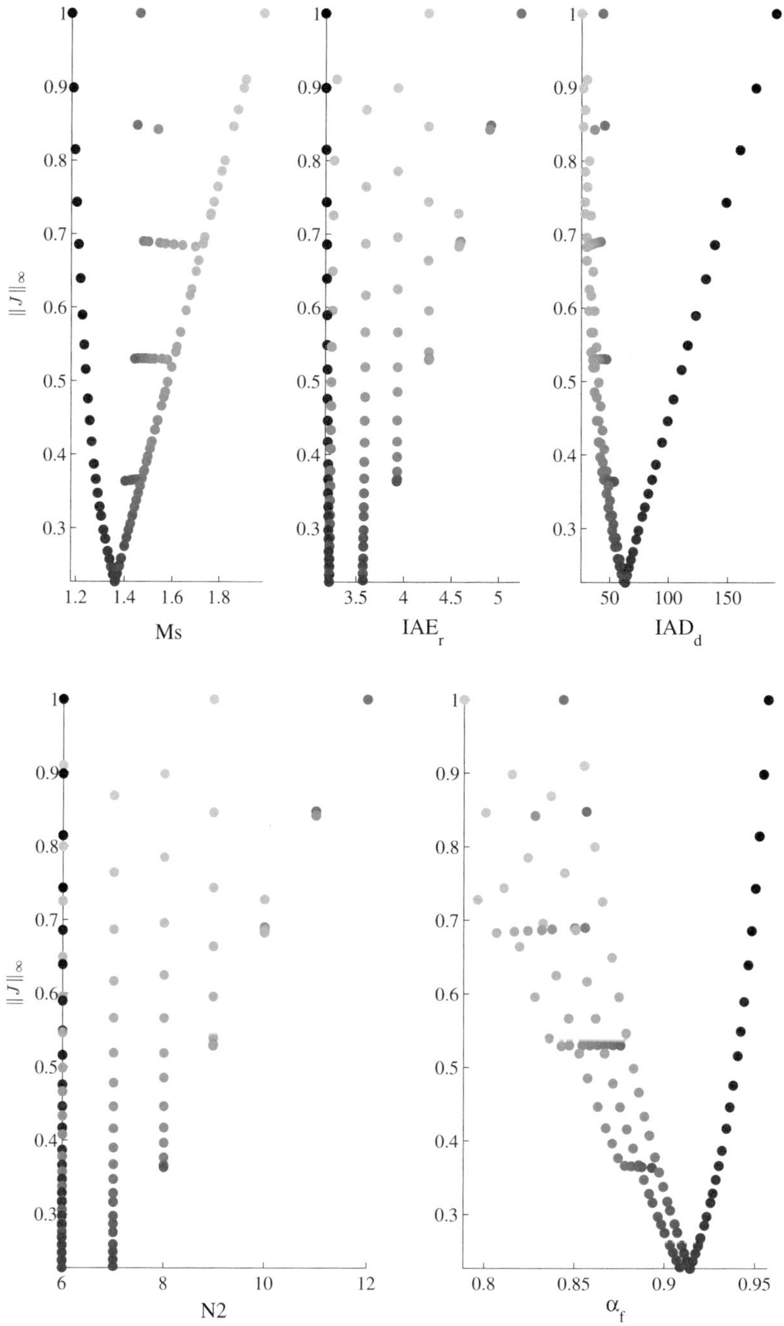

Fig. 6.13 Level diagrams for $\boldsymbol{J}(\boldsymbol{\Theta}_{GPC})$ and $\boldsymbol{\Theta}_{GPC}$. $\|J\|_\infty$ is used for y-axis synchronization

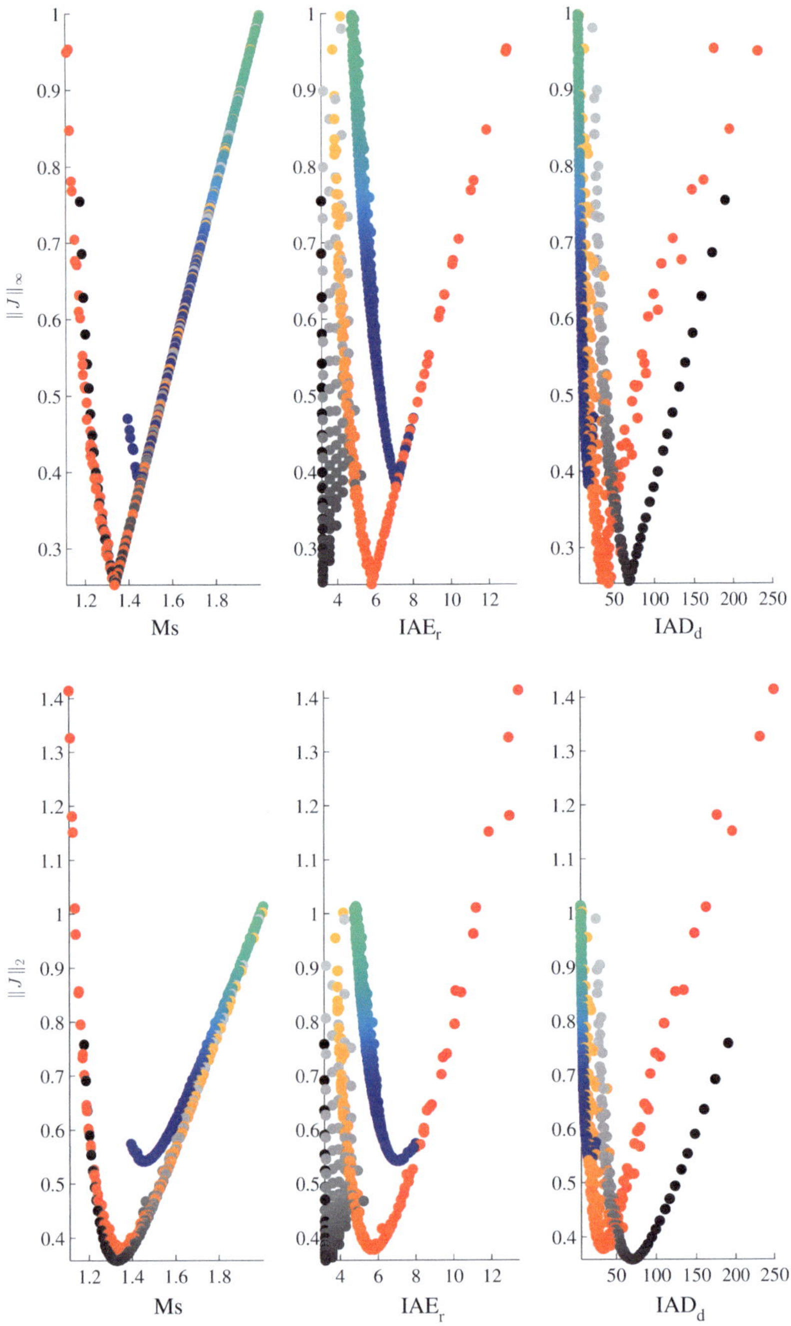

Fig. 6.14 Level Diagram for $\boldsymbol{J}(\boldsymbol{\Theta}_{PI})$, $\boldsymbol{J}(\boldsymbol{\Theta}_{PID})$, $\boldsymbol{J}(\boldsymbol{\Theta}_{GPC})$. Above $\|J\|_\infty$ used for y-axis synchronization. Below $\|J\|_2$ is used

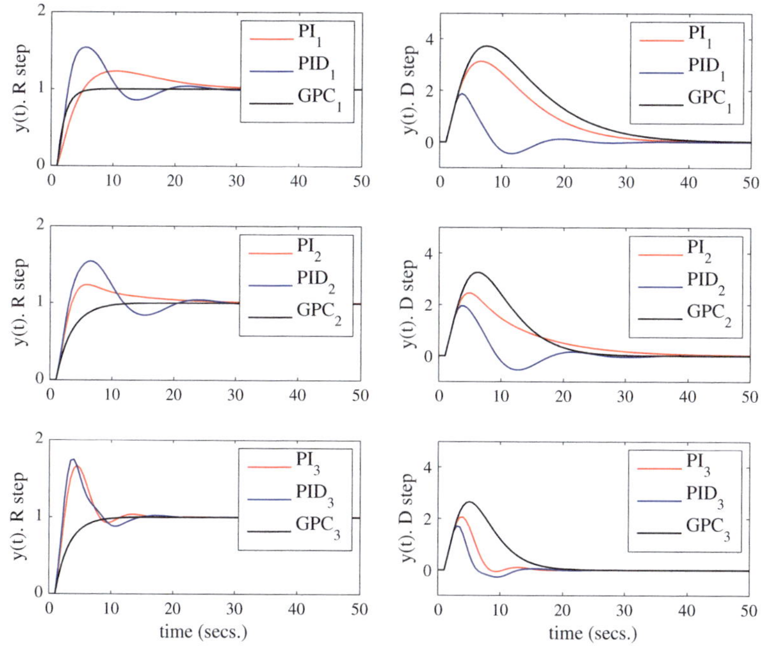

Fig. 6.15 Output $y(t)$ for an unitary step in $r(t)$ at $t = 0$ (*left*). Output $y(t)$ for an unitary step in $d(t)$ at $t = 0$ (*right*)

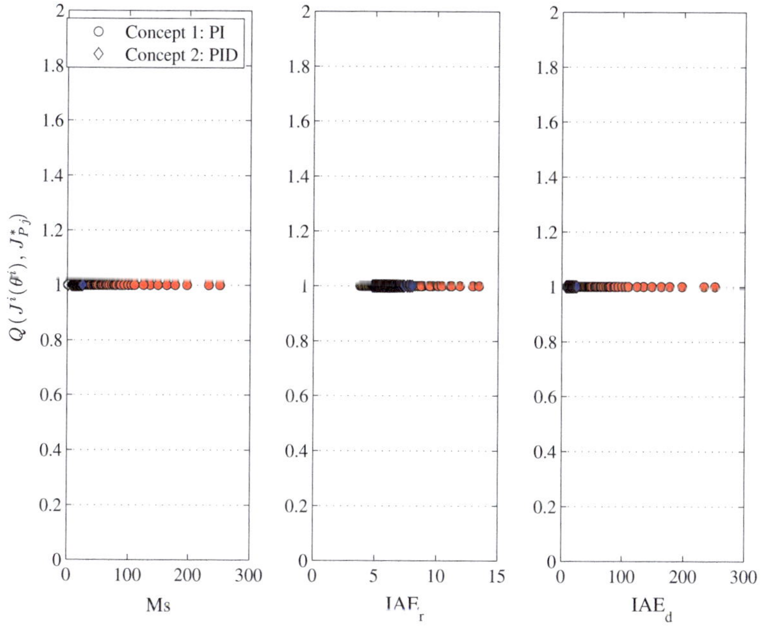

Fig. 6.16 Comparison of PI and PID concepts by using LD and Q indicator for 3D MOP

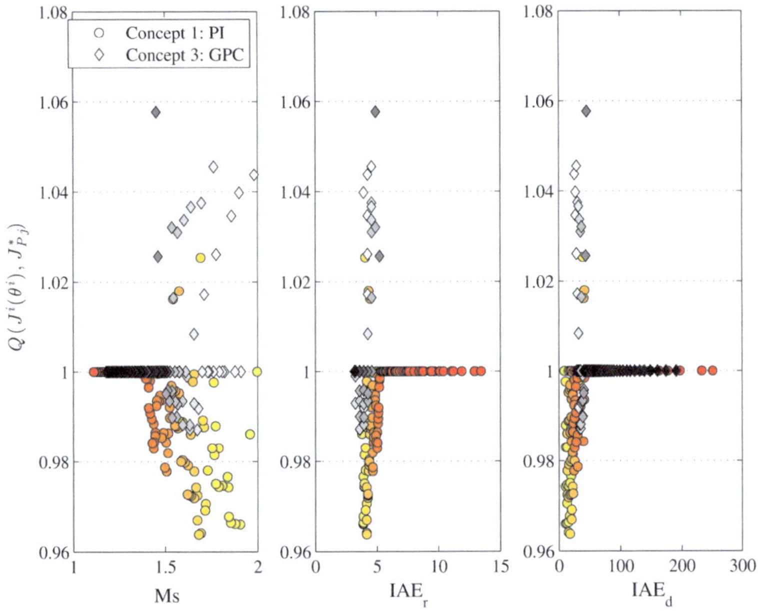

Fig. 6.17 Comparison of PI and GPC concepts by using LD and Q indicator for 3D MOP

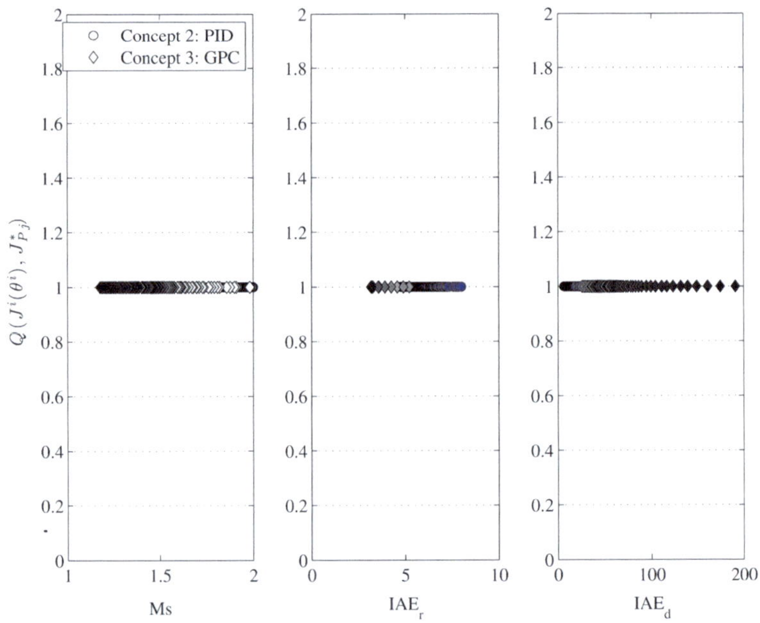

Fig. 6.18 Comparison of PID and GPC concepts by using LD and Q indicator for 3D MOP

6.4 Conclusions

In this chapter, three different controller structures (1-DOF PI, 2-DOF PID and GPC) have been compared under a MOOD approach. Since PI controller only has two parameters to tune, the same number of parameters have been tuned for the rest for a fair comparison. Under these circumstances the example illustrates how using several MO tools when the designer has available different alternatives to solve the problem at hand.

First, a 2D MOP is presented, where robustness and disturbance rejection have been used as objectives. The results show that the GPC controllers (with only two parameters tuned) do not manage well load disturbance and therefore they are dominated by PI and PID controllers. Comparing PI and PID, it has been concluded that depending on the desired degree of robustness it is more appropriate to choose PI or PID controller. For a higher degree of robustness is more appropriate PI than PID, whilst if a lower degree of robustness is needed, PID better reject load disturbances than PIs. Anyway the difference are not quite important and the final decision is adopted according to the controller complexity. As a final remark, for this two objective case, it is possible to analyse the results by using a 2D plot, however LD and quality indicator Q have been used to illustrate their use.

On the other hand, the last example was extended adding a third objective where set-point tracking performance is taken into account. Now a 3D plot is not able to show adequately the results and it is very difficult to compare different controllers. So LD is used for that purpose. Using this tool you can conclude GPC controller presents better performance in set-point tracking than PI or PID ones. Making use of indicator Q, you can conclude also that no design concept dominates another. Therefore the DM task is harder than in the 2D MOP and some preferences have to be considered to obtain the final solution. If set-point tracking results more relevant, DM will select GPC controllers, however if disturbance rejection is a priority, PI controllers are more convenient (since they are simpler than PID and present good balance between robustness and disturbance rejection).

In conclusion, when many objectives have to be managed, having as many tools as possible to compare control alternatives and to supply the DM with extra information is valuable for the final controller selection.

References

1. Åström K, Hägglund T (2001) The future of PID control. Control Eng Pract 9(11):1163–1175
2. Camacho E, Bordons C (1999) Model predictive control. Springer
3. Clarke D, Mohtadi C, Tuffs P (1987) Generalized predictive control-Part I. Automatica 23(2):137–148

4. Clarke D, Mohtadi C, Tuffs P (1987) Generalized predictive control-Part II. Extensions and interpretations. Automatica 23(2):149–160
5. Soeterboek R (1992) Predictive control. A unified approach. Prentice Hall
6. Ziegler JG, Nichols NB (1972) Optimum settings for automatic controllers. Trans ASME 64:759–768

Part III
Benchmarking

This part is devoted to using the multiobjective optimization design (MOOD) procedure in several well known control engineering benchmarks. The aim of this part is twofold: on the one hand evaluating the usefulness of the MOOD procedure in control engineering problem solving; on the other hand, presenting and stating multiobjective optimization versions for such benchmarks, in order to provide to the soft computing community a test-bench to compare multiobjective algorithms and decision making procedures.

Chapter 7
The ACC'1990 Control Benchmark: A Two-Mass-Spring System

Abstract In this chapter, controllers with different complexity for the control benchmark proposed in 1990 at the American Control Conference will be tuned using a multiobjective optimization design procedure. The aim of this chapter is two fold: on the one hand, to evaluate the overall performance of two different controller structures on such a benchmark by means of a design concepts comparison; on the other hand, to state MOP in order to have a more reliable measure of the expected controller's performance.

7.1 Introduction

The robust control benchmark of the American Control Conference (ACC) from 1990 [15] is a popular control problem used in different instances to test different control structures. In the ACC of 1992, several solutions where presented [2–4, 6, 13, 16] and compared [14]. More recently, evolutionary algorithms [7, 8] and MOOD procedures [1, 11] have been used in order to tune different controllers.

While some requirements were provided in the original benchmark [15], performance evaluation in 1992 consisted of a Montecarlo analysis on the risk of failures, regarding settling times and control actions. The aim of this analysis is to enhance the controller's evaluation performance, in order to get a more *reliable* idea (measure) about its performance facing different scenarios. Such scenarios in the benchmark were related with uncertainties in the nominal model.

Since it might be important for the designer to evaluate such reliability, it could be included in the optimization stage, in order to actively seek solutions which minimizes the desired performance in such a Montecarlo analysis. In this case, we are dealing with a reliability-based optimization instance (RBDO) [5].

In this chapter, we will include such design objectives in the MOOD procedure for the ACC-1990 robust control benchmark.

G. Reynoso Meza et al., *Controller Tuning with Evolutionary Multiobjective Optimization*, Intelligent Systems, Control and Automation: Science and Engineering 85, DOI 10.1007/978-3-319-41301-3_7

7.2 Benchmark Setup: ACC Control Problem

The ACC-1990 robust control benchmark [15] consisted in a two-mass-spring system (see Fig. 7.1). It represents a generic model of an uncertain dynamical system with one vibration mode and a rigid body. This system can be modelled as:

$$
\begin{aligned}
\begin{bmatrix} \dot{x}_1 \\ \dot{x}_2 \\ \dot{x}_3 \\ \dot{x}_4 \end{bmatrix} &= \begin{bmatrix} 0 & 0 & 1 & 0 \\ 0 & 0 & 0 & 1 \\ -\frac{k}{m_1} & \frac{k}{m_1} & 0 & 0 \\ \frac{k}{m_2} & -\frac{k}{m_2} & 0 & 0 \end{bmatrix} \begin{bmatrix} x_1 \\ x_2 \\ x_3 \\ x_4 \end{bmatrix} \\
&+ \begin{bmatrix} 0 \\ 0 \\ \frac{1}{m_1} \\ 0 \end{bmatrix} (u + w_1) + \begin{bmatrix} 0 \\ 0 \\ 0 \\ \frac{1}{m_2} \end{bmatrix} w_2 \\
y &= x_2 + v \\
z &= x_2
\end{aligned}
\tag{7.1}
$$

where x_1 and x_2 are the positions of body 1 and 2, respectively; x_3 and x_4 their velocities; u the control action on body 1; y the measured output; w_1, w_2 the plant disturbances; v the sensor noise; z the output to be controlled.

Although in the original benchmark three design problems were stated, just the first one will be used. This problem is devoted to design a linear feedback controller with the following properties:

1. Closed loop must be stable for $m_1 = m_2 = 1$ and $0.5 < k < 2.0$.
2. For a unit impulse $w_2(t)$ at $t = 0$ the settling time should be 15 s. for the nominal model ($k = 1$).
3. Reasonable noise response (designer's choice).
4. Reasonable performance/stability robustness.
5. Minimize control effort.
6. Minimize controller complexity.

In fact the MOP used is a variation of the original one trying to add reliability to the final design. Next, we will state a MOOD procedure for the first instance of such benchmark (Fig. 7.2).

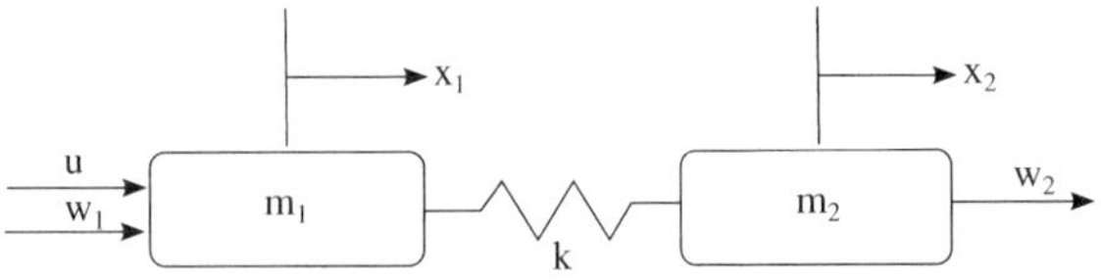

Fig. 7.1 The ACC-1990 robust control benchmark

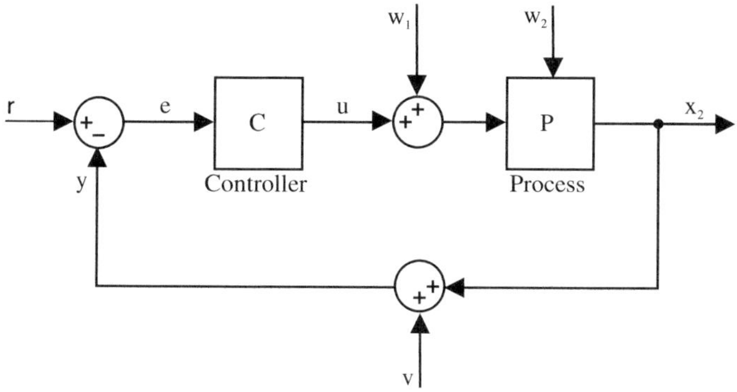

Fig. 7.2 Control loop

7.3 The MOOD Approach

A MOP will be stated trying to add reliability to the final design. In [14] the analysis of the proposed controllers consisted in a Montecarlo analysis on risk failures regarding settling time and control effort. For such purposes, a set Φ of 51 plants, following a normal distribution on $k = [0.5, 2.0]$ was defined. According to this, the following design objectives are defined:

$J_1(\boldsymbol{\theta})$: mean settling time ς_{mean}, in seconds, for the set Φ of 51 different plants, where k follow a uniform distribution between interval [0.5, 2.0]. That is:

$$\begin{aligned} J_1(\boldsymbol{\theta}) &= \varsigma_{mean} = mean(\boldsymbol{\varsigma}) \\ \varsigma_i &= J_{t_{98\%}}(\boldsymbol{\theta}, \phi_i), \forall \phi_i \in \Phi. \end{aligned} \tag{7.2}$$

$J_2(\boldsymbol{\theta})$: maximum settling time ς_{max}, in seconds, for the set Φ of different plants, where k follow a uniform distribution between interval [0.5, 2.0]. That is:

$$\begin{aligned} J_2(\boldsymbol{\theta}) &= \varsigma_{max} - \max(\boldsymbol{\varsigma}) \\ \varsigma_i &= J_{t_{98\%}}(\boldsymbol{\theta}, \phi_i), \forall \phi_i \in \Phi. \end{aligned} \tag{7.3}$$

$J_3(\boldsymbol{\theta})$: maximum control effort u_{max}, in units of u, for the set Φ of different plants, where k follow a uniform distribution between intervals [0.5, 2.0]. That is:

$$\begin{aligned} J_3(\boldsymbol{\theta}) &= u_{max} = \max(\boldsymbol{u}) \\ u_i &= J_{maxU}(\boldsymbol{\theta}, \phi_i), \forall \phi_i \in \Phi. \end{aligned} \tag{7.4}$$

Two different controller structures $C_1(s)$ and $C_2(s)$ will be evaluated and compared (providing a design concepts comparison between two controllers of different complexity).

$$C_1(s) = \frac{\theta_1 s^2 + \theta_2 s + \theta_3}{s^3 + \theta_4 s^2 + \theta_5 s + \theta_6}, \tag{7.5}$$

$$C_2(s) = \frac{\theta_1 s^3 + \theta_2 s^2 + \theta_3 s + \theta_4}{s^4 + \theta_5 s^3 + \theta_6 s^2 + \theta_7 s + \theta_8}. \tag{7.6}$$

According to the benchmark definitions and evaluations, $J_2(\boldsymbol{\theta}) = \varsigma_{max} < 15$ and $J_3(\boldsymbol{\theta}) = u_{max} < 1.0$. Nevertheless, after reviewing the reported results [14], fulfilling the requirement $\varsigma_{max} < 15$ is quite difficult. Therefore, the following MOO statement is defined:

$$\min_{\boldsymbol{\theta}} \boldsymbol{J}(\boldsymbol{\theta}) = [J_1(\boldsymbol{\theta}), J_2(\boldsymbol{\theta}), J_3(\boldsymbol{\theta})] \tag{7.7}$$

subject to:

$$-20 \leq \theta_i \leq 20 \tag{7.8}$$

$$J_1(\boldsymbol{\theta}) = \varsigma_{mean} < 15 \tag{7.9}$$

$$J_2(\boldsymbol{\theta}) = \varsigma_{max} < 30 \tag{7.10}$$

$$J_3(\boldsymbol{\theta}) = u_{max} < 1.0 \tag{7.11}$$

$$\text{Stable in closed loop for nominal model.} \tag{7.12}$$

Since the MOP stated has just 3 objectives and simple pertinency requirements, the sp-MODE algorithm [12] will be used with the parameters depicted in Table 7.1 (details at EMO process section in Chap. 3). Therefore, two different Pareto Set approximations $\boldsymbol{\Theta}^*_{P_1}$, $\boldsymbol{\Theta}^*_{P_2}$ will be calculated, each one corresponding to a design concept ($C_1(s)$ and $C_2(s)$ respectively).

In Fig. 7.3 a design concepts comparison using LD [10] (see details about the comparison tools at the MCDM stage section in Chap. 3). As it can be seen, concept $C_2(s)$ covers a wide range of values; furthermore, some trade-off regions are not accessible for concept $C_1(s)$. For instance, LD for design objectives $J_1(\boldsymbol{\theta})$ and $J_3(\boldsymbol{\theta})$ show how concept $C_2(s)$ reaches a trade-off region with better $J_1(\boldsymbol{\theta})$, but worse $J_3(\boldsymbol{\theta})$ providing a better performance at the expense of bigger values of control action when compared with $C_1(s)$. It is possible to appreciate this at J_1 in the range

Table 7.1 Parameters used for sp-MODE. Further details in Chap. 3

		Parameter	Value
sp-MODE	DE	**Evolutionary mechanism**	
		F (Scaling factor)	0.5
		Cr (Crossover rate)	0.9
		N_p (Population)	20
		Pruning mechanism	
		β^{ϵ} (Arcs)	15

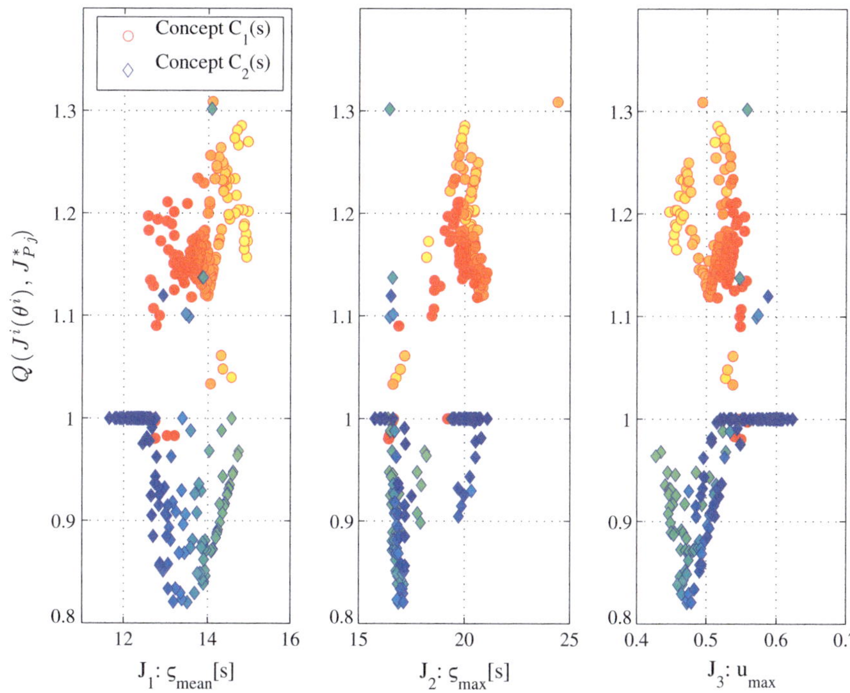

Fig. 7.3 Design concepts comparison for controller structures $C_1(s)$ and $C_2(s)$

[11.5, 12.5] and $J_3 > 0.55$ (approximately) since values of the quality indicator $Q(\boldsymbol{J}^i(\boldsymbol{\theta}^i), \boldsymbol{J}^*_{P_j})$ for concept $C_2(s)$ are 1 and simultaneously there are no solutions of concept $C_1(s)$. Nevertheless, their difference is evident in $J_2(\boldsymbol{\theta})$, where concept $C_1(s)$ has a tendency to allow a higher maximum settling time when compared with concept $C_2(s)$. Besides, the *exclusive* trade-off region commented before for design concept $C_2(s)$ belongs to the regions on $J_2(\boldsymbol{\theta})$ with the higher or the lower values of the maximum settling time attained (points with a quality indicator of 1 at extremes).

In an overall picture, concept $C_2(s)$ dominates concept $C_1(s)$, which is shown in this visualization paradigm, when solutions with quality indicator over 1 are dominated by solutions below 1. Notice that it is possible to appreciate that Pareto Front approximation $\boldsymbol{J}^*_{P_1}$ (concept $C_1(s)$) is above 1 and the Pareto Front approximation $\boldsymbol{J}^*_{P_2}$ (concept $C_2(s)$) is below. From the engineering point of view, you can say that it is justifiable to use a controller with higher complexity (number of poles and zeros) only if it is important for this application to assure a maximum settling time below 20 s; that is, design alternatives with a trade-off not provided by the concept $C_2(s)$. If this fact is not justifiable, therefore this control application can be managed with the controller of lower complexity.

Concerning the MCDM stage for each design concept, Pareto Front and Set approximations $\boldsymbol{\Theta}^*_{P_1}$, $\boldsymbol{J}^*_{P_1}$ and $\boldsymbol{\Theta}^*_{P21}$, $\boldsymbol{J}^*_{P_2}$ are depicted respectively in Figs. 7.4 and 7.5. After an analysis on such approximations, two controllers $\boldsymbol{\theta}_{C1_{DM}}$ and $\boldsymbol{\theta}_{C2_{DM}}$ are selected for further control tests (marked with ★ in figures):

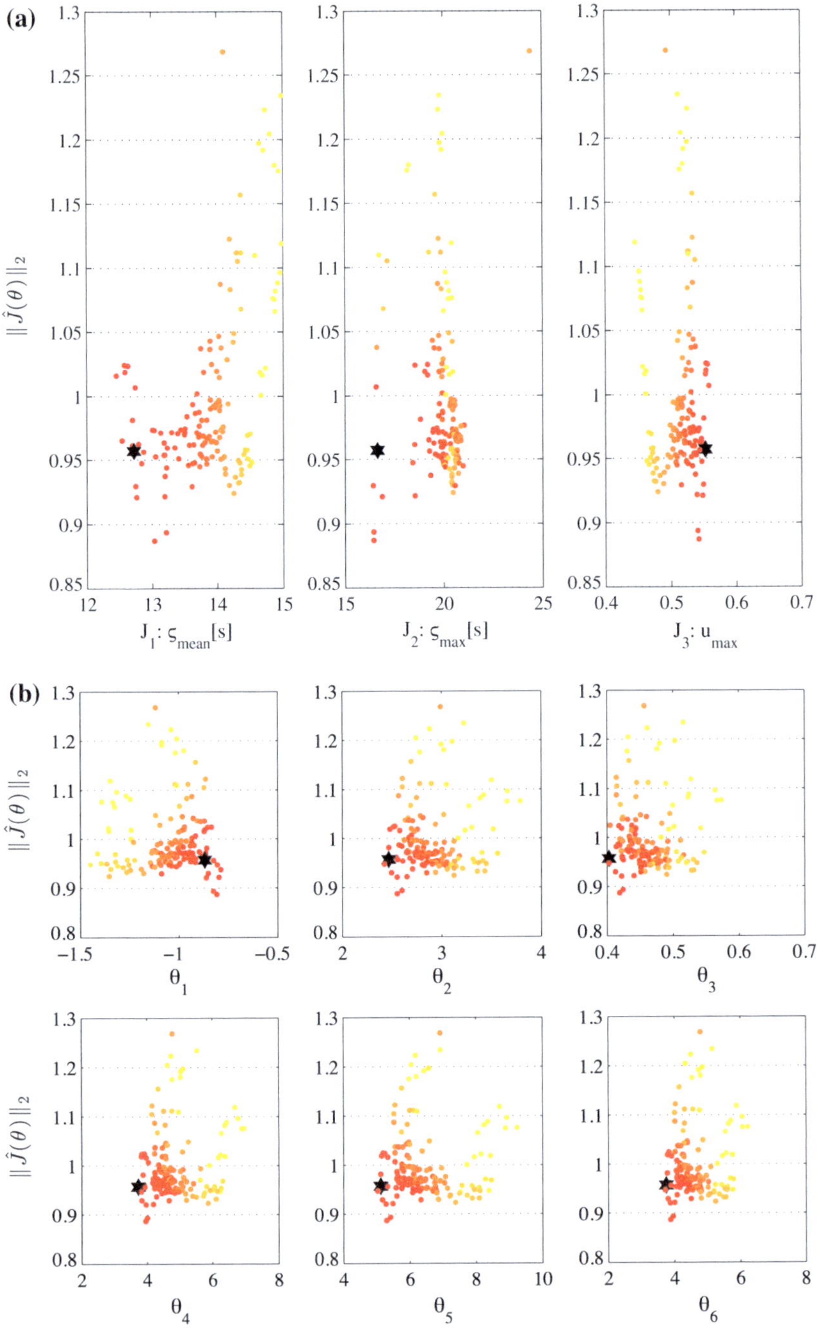

Fig. 7.4 Pareto front and set approximated with the controller structure $C_1(s)$ (design concept 1). ★ remarks the $\boldsymbol{\theta}_{C1_{DM}}$ controller. **a** Pareto Front. **b** Pareto Set

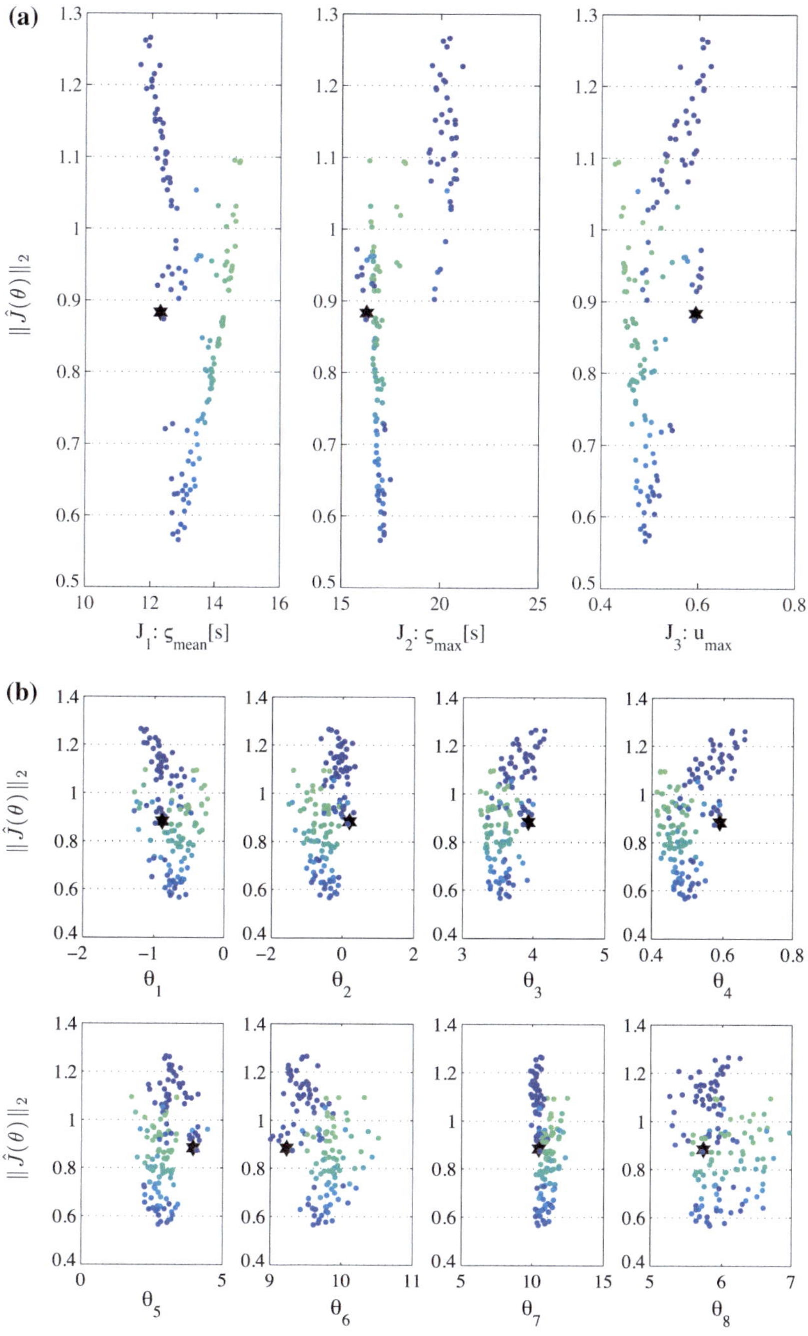

Fig. 7.5 Pareto front and set approximated with the controller structure $C_2(s)$ (design concept 2). ★ remarks the $\boldsymbol{\theta}_{C2_{DM}}$ controller. **a** Pareto Front. **b** Pareto Set

$$C_{1_{DM}}(s) = \frac{-0.8658s^2 + 2.4643s + 0.4031}{s^3 + 3.7307s^2 + 5.1249s + 3.7481}, \tag{7.13}$$

$$C_{2_{DM}}(s) = \frac{-0.8885s^3 + 0.1872s^2 + 3.9257s + 0.5896}{s^4 + 3.9526s^3 + 9.2342s^2 + 10.4753s + 5.7370}. \tag{7.14}$$

The general criteria to select such controllers was to achieve a good trade-off between settling time and its variation (measured with the maximum settling time reported), since in all cases, controllers fulfill the minimum control effort constraint.

7.4 Control Tests

Risk failure, as in [14], will be calculated for controllers $C_{1_{DM}}(s)$ and $C_{2_{DM}}(s)$. For that purpose, 20,000 different plants will be used to evaluate their performance. Notice that in the previously MOP statement only 51 different plants were used for the sake of simplicity and to avoid an impractical approach with such computational burden to perform an optimization stage. Risk of failure is related with having a settling time bigger than 15 s and a maximum control effort bigger than 1. For comparison purposes, the following reference controllers (from [9, 14] respectively) are also considered:

$$C_{1_R}(s) = \frac{-12.5000s^2 + 12.8375s + 3.1211}{s^3 + 21.8124s^2 + 26.4400s + 30.1605}, \tag{7.15}$$

$$C_{2_R}(s) = \frac{-2.1300s^3 - 5.3270s^2 + 6.2730s + 1.0150}{s^4 + 4.6800s^3 + 12.9400s^2 + 18.3600s + 12.6800}. \tag{7.16}$$

Table 7.2 shows the results of risk failure; also Fig. 7.6a, b depict time responses for the set of uncertainties Φ used in the optimization stage. Both controllers tuned by the MOOD procedure, achieved a low risk of failure. In the case of maximum control effort, the reference controllers in any case were below 1, nevertheless, this is in exchange of having a 100 % of risk failure in settling time for low complexity structure or around 80 % with a more complex structure. It is worthwhile to say that such controllers have a disadvantage, since their tuning procedure did not take into account such a kind of Montecarlo analysis for a reliable measure of their performances.

Table 7.2 Risk of failure for settling time and maximum control effort

Controller	Settling time (%)	Control effort (%)
$C_{1_{DM}}(s)$	20.41	2.34
$C_{1_R}(s)$	100.0	0.00
$C_{2_{DM}}(s)$	11.22	2.13
$C_{2_R}(s)$	79.46	0.00

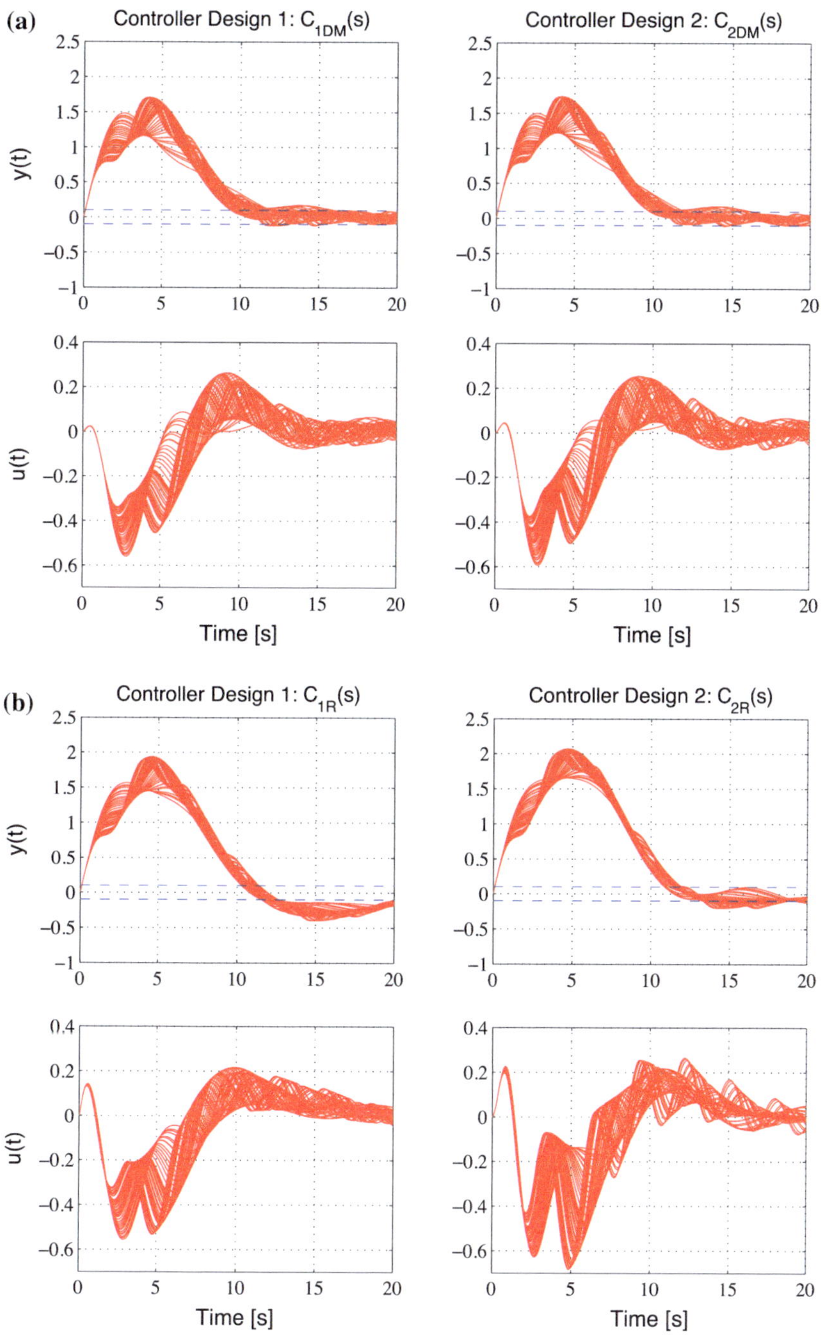

Fig. 7.6 Time responses comparison among controllers (51 random models tested for each controller). **a** $C1_{DM}(s)$ and $C2_{DM}(s)$. **b** $C1_R(s)$ and $C2_R(s)$

7.5 Conclusions

In this chapter, Pareto Fronts for two controller (with different structure) were approximated, in order to have an overall comparison (instead of point by point) of the achievable trade-off among conflicting objectives. With such comparison, it was possible to identify strengths of one controller structure (the complex) over the other (the simple one) in such a way that the designer is able to ponder if such improvement on performance compensates using one structure over the other.

For this benchmark, the MOP statement defined is more in concordance with the expected performance and risk of failure, via a Montecarlo analysis. This kind of MOP are based on reliability, where optimization approaches seek to guarantee a given performance, when dealing, as in this case, with inaccuracies in the model. The improvement over other controllers reported in the literature lies on the fact that, in this case, the MOP was minding such evaluation criteria, used at the end of the process by the DM. That is, the MOOD procedure using EMO enables us to define a more meaningful MOP statement, closer to the DM preferences and desired performance.

References

1. Blasco X, Herrero J, Sanchis J, Martínez M (2008) A new graphical visualization of n-dimensional pareto front for decision-making in multiobjective optimization. Inf Sci 178(20):3908–3924
2. Byrns Jr EV, Calise AJ (1990) Fixed order dynamic compensation for the h2/hinf benchmark problem. In: American control conference, 1990. IEEE, pp 963–965
3. Chiang R, Safonov M (1990) H robust control synthesis for an undamped, non-colocated spring-mass system. In: American control conference. IEEE, pp 966–967
4. Collins E, Bernstein D (1990) Robust control design for a benchmark problem using a structured covariance approach. In: American control conference, no 27, pp 970–971
5. Frangopol DM, Maute K (2003) Life-cycle reliability-based optimization of civil and aerospace structures. Comput Struct 81(7):397–410
6. Ly U-L (1990) Robust control design using nonlinear constrained optimization. In: American control conference, 1990. IEEE, pp 968–969
7. Martínez M, Sanchis J, Blasco X (2006) Multiobjective controller design handling human preferences. Eng Appl Artif Intell 19:927–938
8. Martínez MA, Sanchis J, Blasco X (2006) Algoritmos genéticos aplicados al diseño de controladores robustos. RIAII 3(1):39–51
9. Messac A, Wilsont B (1998) Physical programming for computational control. AIAA J 36(2):219–226
10. Reynoso-Meza G, Blasco X, Sanchis J, Herrero JM (2013) Comparison of design concepts in multi-criteria decision-making using level diagrams. Inf Sci 221:124–141
11. Reynoso-Meza G, Sanchis J, Blasco X, García-Nieto S (2014) Physical programming for preference driven evolutionary multi-objective optimization. Appl Soft Comput 24:341–362
12. Reynoso-Meza G, Sanchis J, Blasco X, Martínez M (2010) Design of continuous controllers using a multiobjective differential evolution algorithm with spherical pruning. In: Applications of evolutionary computation. Springer, pp 532–541
13. Rhee I, Speyer JL (1990) Application of a game theoretic controller to a benchmark problem. Am Control Conf 1990:972–973 May

14. Stengel RF, Marrison CI (1992) Robustness of solutions to a benchmark control problem. J Guid Control Dyn 15:1060–1067
15. Wie B, Bernstein DS (1990) A benchmark problem for robust control design. Am Control Conf 1990:961–962
16. Wie B, Bernstein DS (1992) Benchmark problems for robust control design. J Guid Control Dyn 15:1057–1059

Chapter 8
The ABB'2008 Control Benchmark: A Flexible Manipulator

Abstract In this chapter, a digital controller is tuned for the control benchmark proposed in 2008 by the ABB group at the 17th IFAC World Congress via multi-objective optimization. In some instances, a more realistic evaluation of a controller performance is sought, that is, the expected performance of the controller when it will be implemented. For this benchmark, a digital controller with limited control actions is adjusted in order to control the end effector of a robotic arm.

8.1 Introduction

The ABB control benchmark problem [1] is a complete and realistic simplification of a regulatory problem for a manipulator's end effector (IRB6600, ABB©). The aim of the benchmark is to define a controller (with free structure) in order to keep the desired reference (tool position) when dealing with disturbances in torque and end tool. For this benchmark, some specifications were given, in order to state a most reliable performance evaluation of the controller to be implemented. Such specifications are related with the structure of the controller: should be delivered in its digital form, for a sampling rate of 5 ms. Besides, a test is defined and several indicators are aggregated into an AOF to evaluate the overall performance of a given controller.

The evaluation also considers a reliable performance measure since it will be evaluated in a set of different plants, given some uncertainty in the nominal model parameters. Nonetheless, in this case an active search is not practical, due to the computational burden of the model. Therefore, a two stage MOOD procedure will be stated in order to accomplish a suitable design.

8.2 Benchmark Setup: The ABB Control Problem

The ABB control benchmark problem is a complete and realistic simplification of the regulatory problem of a manipulator's end effector (IRB6600, ABB©). Its main simplifications are the following:

G. Reynoso Meza et al., *Controller Tuning with Evolutionary Multiobjective Optimization*, Intelligent Systems, Control and Automation: Science and Engineering 85, DOI 10.1007/978-3-319-41301-3_8

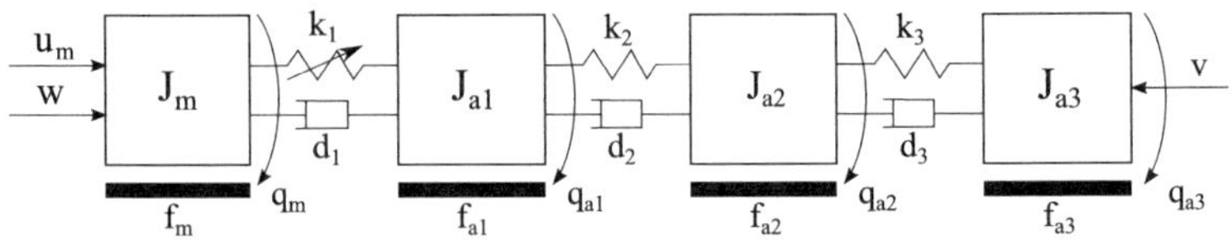

Fig. 8.1 Four masses model for IRB6600

- Just the first axis of the manipulator is considered.
- Its dynamics is modeled with a four masses model (see Fig. 8.1).
- Current is assumed ideal. Torque in the motor is saturated between ±20 Nm.
- Friction effects are considered linear.

According to the Fig. 8.1, the system under consideration is:

$$J_m\ddot{q}_m = u_m + w - f_m\dot{q}_m - \tau_{gear} - d_1(\dot{q}_m - \dot{q}_{a1}) \tag{8.1}$$

$$J_{a1}\ddot{q}_{a1} = -f_{a1}\dot{q}_{a1} + \tau_{gear} + d_1(\dot{q}_m - \dot{q}_{a1}) - k_2(q_{a1} - q_{a2}) - d_2(\dot{q}_{a1} - \dot{q}_{a2}) \tag{8.2}$$

$$J_{a2}\ddot{q}_{a2} = -f_{a2}\dot{q}_{a2} + k_2(q_{a1} - q_{a2}) + d_2(\dot{q}_{a1} - \dot{q}_{a2}) - k_3(q_{a2} - q_{a3}) - d_3(\dot{q}_{a2} - \dot{q}_{a3}) \tag{8.3}$$

$$J_{a3}\ddot{q}_{a3} = v - f_{a3}\dot{q}_{a3} + k_3(q_{a2} - q_{a3}) + d_3(\dot{q}_{a2} - \dot{q}_{a3}) \tag{8.4}$$

where J_{a1}, J_{a2} and J_{a3} are the inertia moments of the arm; J_m the inertia of the motor; q_{a1}, q_{a2} and q_{a3} the angles of the three masses; τ_{gear} a nonlinear function of the deflection $q_m - q_{a1}$ (first spring-damper pair) approximated by a piece-wise linear function[1]; d_1, d_2 and d_3 spring linear dampings; k_2, k_3 linear elasticity; z tool position; $f_m, f_{a1}, f_{a2}, f_{a3}$ viscous frictions in the motor and the arm structure, respectively; w and v motor and tool torque disturbances, respectively and finally q_m represents the motor angle. The challenge is to control the tool position z:

$$z = \frac{l_1 q_{a1} + l_2 q_{a2} + l_3 q_{a3}}{r} \tag{8.5}$$

where r is the gear-radius and l_1, l_2 and l_3 the distance between masses and the tool (Fig. 8.2).

The benchmark challenge was stated in three phases:

- For a given nominal model, defined in Table 8.1 (hereafter *Nom*).
- For a set of models, with small variations in their physical parameters (hereafter Set-1).
- For a set of models, with significant variations in their physical parameters (hereafter Set-2).

[1]Five segments, but only three are given: $k_{1,high}$, $k_{1,low}$, $k_{1,pos}$.

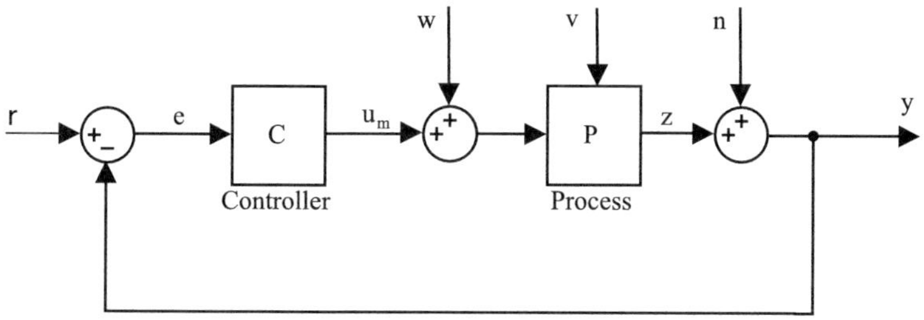

Fig. 8.2 Control loop

Table 8.1 Nominal parameters values

Parameter	Value	Unit
J_m	$5e-3$	kg m^2
J_{a1}	$2e-3$	kg m^2
J_{a2}	$2e-2$	kg m^2
J_{a3}	$2e-2$	kg m^2
$k_{1,high}$	100	Nm/rad
$k_{1,low}$	16.7	Nm/rad
$k_{1,pos}$	$64e-3$	rad
k_2	110	Nm/rad
k_3	80	Nm/rad
d_1	$8e-2$	Nm s/rad
d_2	$6e-2$	Nm s/rad
d_3	$8e-2$	Nm s/rad
f_m	$6e-3$	Nm s/rad
f_{a1}	$1e-3$	Nm s/rad
f_{a2}	$1e-3$	Nm s/rad
f_{a3}	$1e-3$	Nm s/rad
r	220	
l_1	20	mm
l_2	600	mm
l_3	1530	mm
T_d	$5e-3$	s

Being an industrial application, time-domain measures and constraints are used to evaluate the performance of a given controller (such measures are meaningful and easy to understand):

- Peak to peak error $(e_1, \ldots, e_8)$ [mm].
- Settling times $(t_S^1, \ldots, t_S^4)$ [s].
- Maximum torque (T_{MAX}) [Nm].

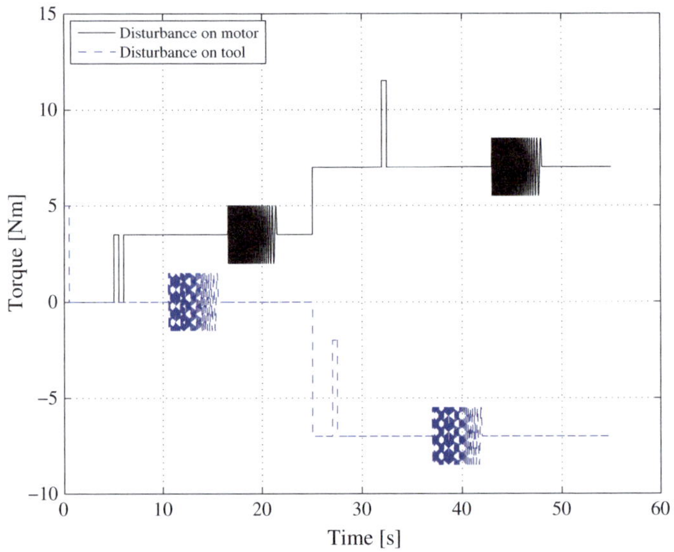

Fig. 8.3 Torque disturbance for control evaluation

- Adjusted rms value (T_{rms}) [Nm].
- Torque noise, peak to peak (T_{NOISE}) [Nm].

The test used for controller evaluation is depicted in Fig. 8.3. Design requirements are as follows:

1. Settling times for nominal model and Set-1: $t_S^i < 3$ s with and error band of ± 0.1 mm.
2. Settling times for Set-2: $t_S^i < 4$ s with and error band of ± 0.3 mm.
3. $T_{NOISE} < 5$ [Nm].
4. Stability when increasing loop gain by 2.5 and when increasing delay time to 2 [ms].

An AOF was defined in order to evaluate the overall performance of a controller $\boldsymbol{C}$, for each benchmark phase:

$$V_{Nom}(\boldsymbol{C}) = \sum_{i=1}^{15} \alpha_i f_i(\boldsymbol{C}) \tag{8.6}$$

$$V_{Set-1}(\boldsymbol{C}) = \sum_{i=1}^{15} \alpha_i \max_{m \in SET-1} (f_i(\boldsymbol{C})) \tag{8.7}$$

$$V_{Set-2}(\boldsymbol{C}) = \sum_{i=1}^{15} \alpha_i \max_{m \in SET-2} (f_i(\boldsymbol{C})) \tag{8.8}$$

where

$$\begin{aligned}\boldsymbol{f}(\boldsymbol{C}) &= [e_1, e_2, e_3, e_4, e_5, e_6, e_7, e_8, t_s^1, t_s^2, t_s^3, t_s^4, T_{NOISE}, T_{MAX}, T_{RMS}]\\ \boldsymbol{\alpha} &= [0.7, 1.4, 1.4, 2.8, 0.7, 1.4, 1.4, 2.8, 2.8, 2.8, 2.8, 2.8, 1.4, 1.4, 3.5].\end{aligned}$$

Finally, a global index for a given controller $\boldsymbol{C}$ is provided by another AOF with linear weighting, using the three phases of the benchmark:

$$V(\boldsymbol{C}) = \beta_1 V_{Nom}(\boldsymbol{C}) + \beta_2 V_{Set-1}(\boldsymbol{C}) + \beta_3 V_{Set-2}(\boldsymbol{C}) \quad (8.9)$$

with $\beta_1 = 0.6$, $\beta_2 = 1.0$ and $\beta_3 = 0.3$.

The next step will be to set two different MOPs to address such a control problem.

8.3 The MOOD Approach

A PID with derivative filter in its parallel form will be used as $\boldsymbol{C}$ for the benchmark.

$$\boldsymbol{C} = K_c + \frac{K_i}{s} + \frac{K_d \cdot s}{F_p \cdot s + 1}. \quad (8.10)$$

This is justified since, even if different methodologies and proposals were submitted, an order reduction from a presented controller to a PID like form was possible, keeping a reasonable performance, according to the benchmark index (8.9).

As commented in the introduction, dealing with uncertainties in the system using an active search approach such as Montecarlo analysis in the optimization stage (as in Chap. 7) is not practical; this is due to the fact that the test platform, with the degree of realism incorporated, might cause a considerable computational burden for such optimization approach. Therefore, a MOP using only the nominal model will be stated including robustness measures, in order to face the system uncertainties. According to this, the following MOP is defined:

$$\begin{aligned}\min_{\boldsymbol{\theta}} \boldsymbol{J}(\boldsymbol{\theta}) &= [J_1(\boldsymbol{\theta}), J_2(\boldsymbol{\theta}), J_3(\boldsymbol{\theta})] \quad (8.11)\\ \boldsymbol{\theta} &= [K_c, K_i, K_d, F_p]\end{aligned}$$

where design objectives are:

$J_1(\boldsymbol{\theta})$: benchmark index $V_{Nom}(\boldsymbol{C})$ for the nominal model; for such purpose, scripts provided by the organizers will be used.

$J_2(\boldsymbol{\theta})$: Maximum value of the sensitivity function, with the nominal model $P_N(s) = \frac{z(s)}{u_m(s)}$ according to Table 8.1,

$$J_2(\boldsymbol{\theta}) = \left\| (1 + P_N(s) C_{\boldsymbol{\theta}}(s))^{-1} \right\|_\infty, \quad (8.12)$$

$J_3(\boldsymbol{\theta})$: Maximum gain G at frequency range between [1, 1000].

$$J_3(\boldsymbol{\theta}) = \|C_{\boldsymbol{\theta}}(jw)\|_{w\in[1,\ 1000]}. \tag{8.13}$$

Using as reference the PID controller provided by the organizers, the pertinency region of the approximated Pareto Front is bounded using the performance of this controller. Finally the MOP statement is:

$$\min_{\boldsymbol{\theta}} \boldsymbol{J}(\boldsymbol{\theta}) = [J_1(\boldsymbol{\theta}), J_2(\boldsymbol{\theta}), J_3(\boldsymbol{\theta})] \tag{8.14}$$
$$\boldsymbol{\theta} = [K_c, K_i, K_d, F_p]$$

subject to:

$$1 \leq K_p \leq 60 \tag{8.15}$$
$$0 \leq K_i \leq 150 \tag{8.16}$$
$$0 \leq K_d \leq 6 \tag{8.17}$$
$$0.01 \leq F_p \leq 1 \tag{8.18}$$
$$J_1(\boldsymbol{\theta}) < 85 \tag{8.19}$$
$$1.1 \leq J_2(\boldsymbol{\theta}) \leq 1.8 \tag{8.20}$$
$$J_3(\boldsymbol{\theta}) \leq 60dB \tag{8.21}$$
$$\text{Stable in closed loop.} \tag{8.22}$$

Since only three design objectives are managed and simple pertinency bounds are defined, the sp-MODE algorithm [2] is ran (with parameters of Table 8.2) obtaining the Pareto Front $\boldsymbol{J}^*_{P_1}$ and set $\boldsymbol{\Theta}^*_{P_1}$ of Fig. 8.4.

After analysing such Pareto Front and performing the MCDM stage with the full benchmark, we can notice that this controller structure can achieve values up to $V_{Nom}(\boldsymbol{C}) = 61$ with the nominal model. Nevertheless, as expected, several of those

Table 8.2 Parameters used for sp-MODE. Further details in Chap. 3

		Parameter	Value
sp-MODE		**Evolutionary mechanism**	
	DE	F (Scaling factor)	0.5
		Cr (Crossover rate)	0.9
		N_p (Population)	50
		Pruning mechanism	
		$\boldsymbol{\beta}^{\epsilon}$ (Arcs)	20

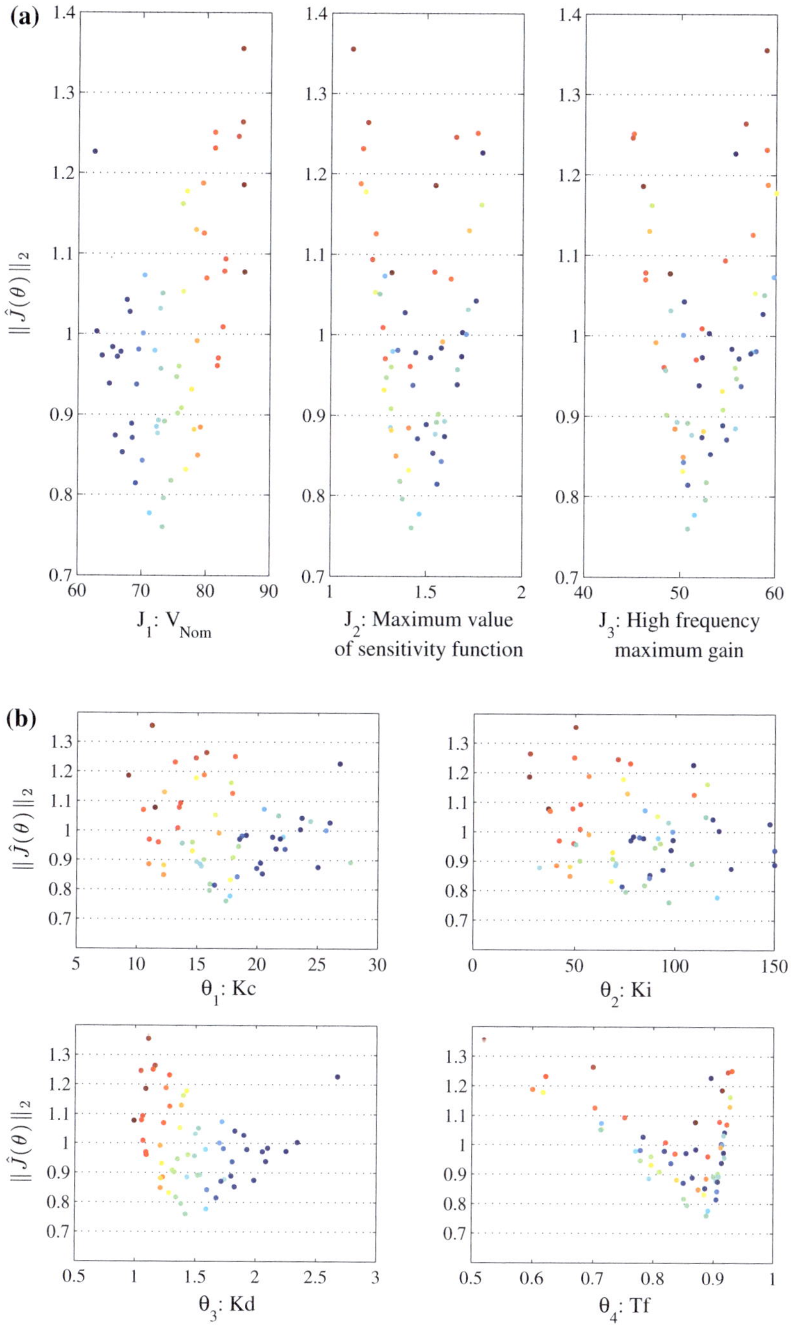

Fig. 8.4 Pareto front $\boldsymbol{J}^*_{P_1}$ and set $\boldsymbol{\Theta}^*_{P_1}$. **a** Pareto front. **b** Pareto set

controllers perform badly when gain or delay are increased or when other models are test (Set-1 and Set-2) given the exchange in robustness, measured with objective $J_2(\boldsymbol{\theta})$. Looking at $\boldsymbol{J}^*_{P_1}$, it is possible to have an idea about the possibilities on this controller structure and it is possible a further refinement of the search process. According to this recently acquired knowledge, a new MOP is stated:

$$\min_{\boldsymbol{\theta}} \boldsymbol{J}(\boldsymbol{\theta}) = [J_1(\boldsymbol{\theta}), J_2(\boldsymbol{\theta}), J_3(\boldsymbol{\theta})] \tag{8.23}$$
$$\boldsymbol{\theta} = [K_c, K_i, K_d, F_p]$$

subject to:

$$1 \leq K_p \leq 60 \tag{8.24}$$
$$0 \leq K_i \leq 150 \tag{8.25}$$
$$0 \leq K_d \leq 6 \tag{8.26}$$
$$0.01 \leq F_p \leq 1 \tag{8.27}$$
$$J_1(\boldsymbol{\theta}) < 65 \tag{8.28}$$
$$J_2(\boldsymbol{\theta}) \leq 1.8 \tag{8.29}$$
$$J_3(\boldsymbol{\theta}) \leq 60[dB] \tag{8.30}$$
$$\pi/Wcp < 10/1000 \tag{8.31}$$
$$\text{Stable in closed loop.} \tag{8.32}$$

where an additional constraint is related to phase margin frequency, Wcp is included in order to guarantee stability when the delay is increased. Furthermore, for this new MOP the EMO process will be used as initial population the suitable controllers from the first Pareto Set $\boldsymbol{\Theta}^*_{P_1}$. This sequential optimization is performed since it might be expected to have little knowledge on the achievable trade-off for a given controller structure. Therefore running one or two additional optimization instances could be helpful in order to refine the search process. After such optimization process, new Pareto Set $\boldsymbol{\Theta}^*_{P_2}$ and Front $\boldsymbol{J}^*_{P_2}$ are approximated (See Fig. 8.5).

After an analysis of the objectives exchange showed by $\boldsymbol{J}^*_{P_2}$ (Fig. 8.5a) a controller is selected (depicted with a ★):

$$\boldsymbol{\theta}_{DM} = [21.9853, 85.2029, 2.3390, 0.9031].$$

This controller has been preferred over the one with the minimum 2-norm, due to its lower values in K_c, K_i (See Fig. 8.5b). Such controller is taken for further control tests.

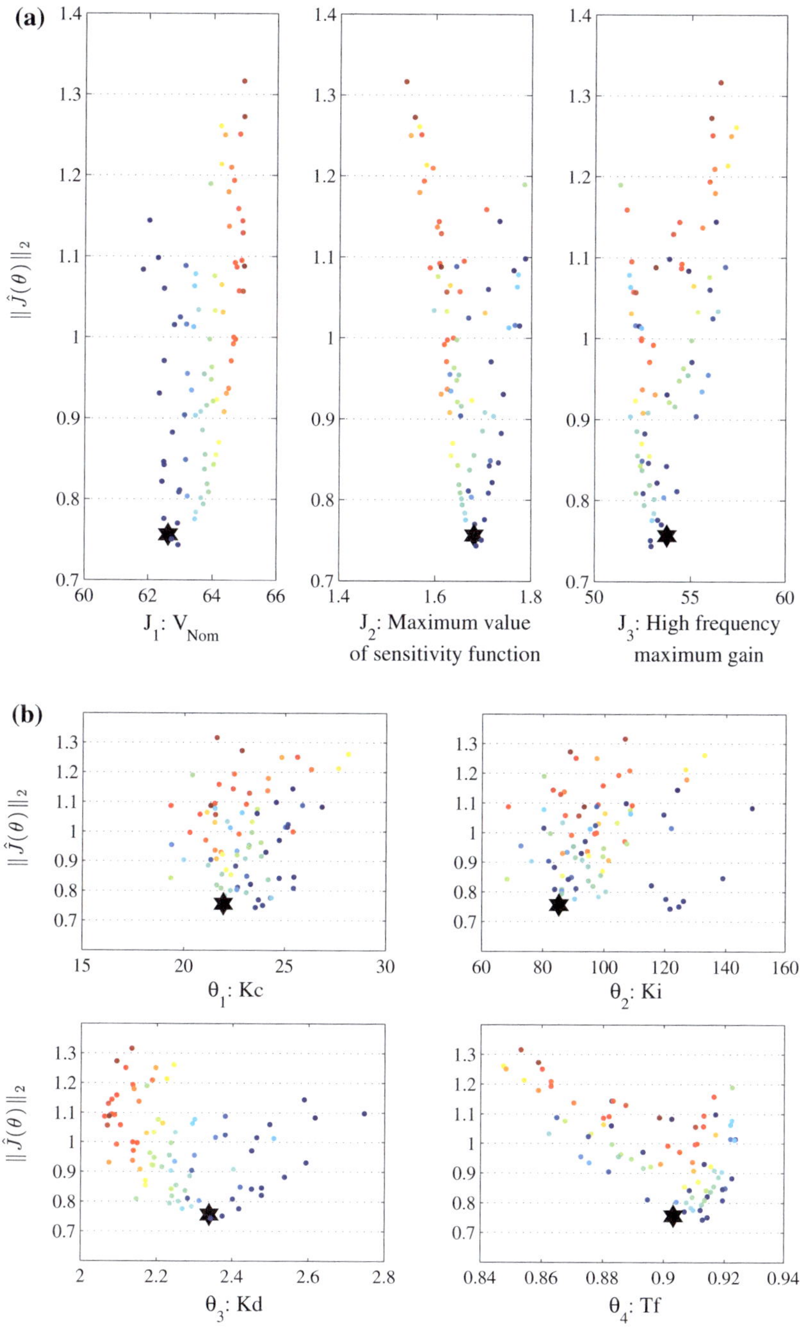

Fig. 8.5 Pareto front $J^*_{P_2}$ and set $\Theta^*_{P_2}$ approximations. **a** Pareto front. **b** Pareto set

8.4 Control Tests

The selected controller has been evaluated with the nominal model and Set-1 and Set-2 of uncertainties, in order to calculate the global index of Eq. 8.9. Time responses are depicted in Fig. 8.6 for the nominal model, in Fig. 8.7 with a gain increase and in Fig. 8.8 with a delay increase (recall design requirement number 3). Notice that the

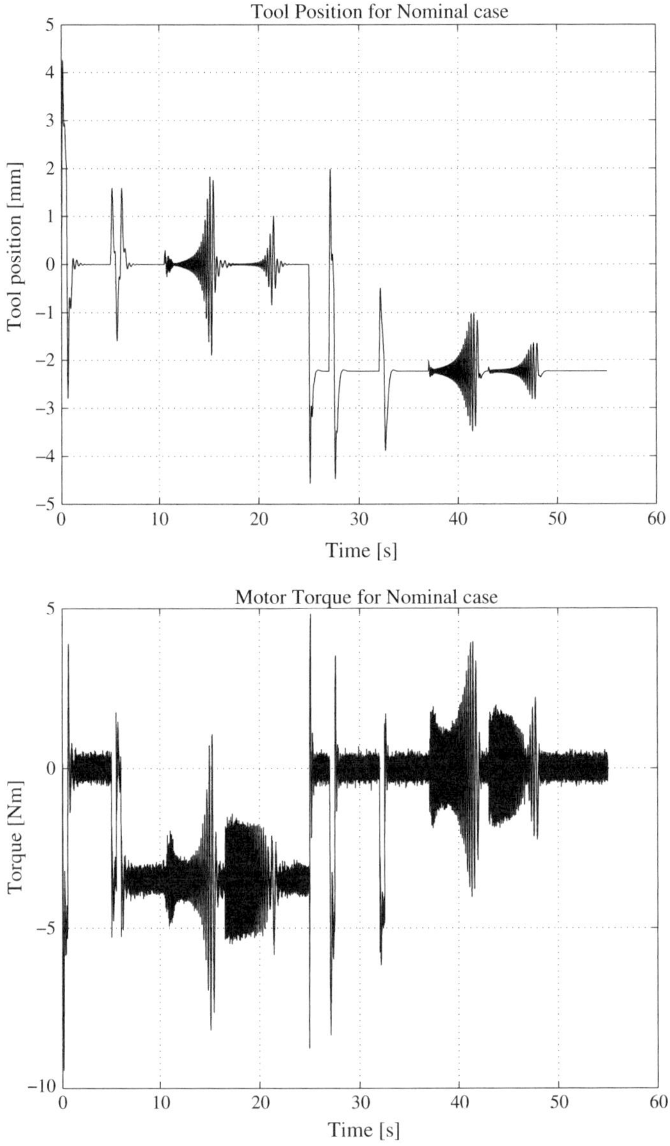

Fig. 8.6 Time response performance of the selected controller $\boldsymbol{\theta}_{DM}$ for the nominal model

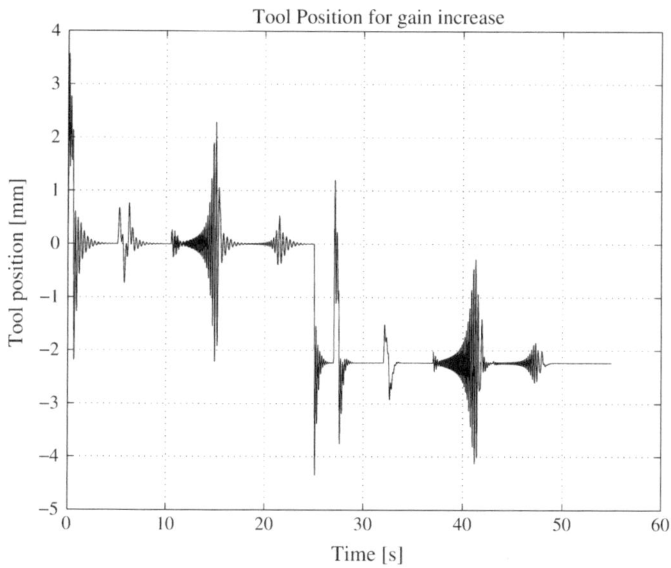

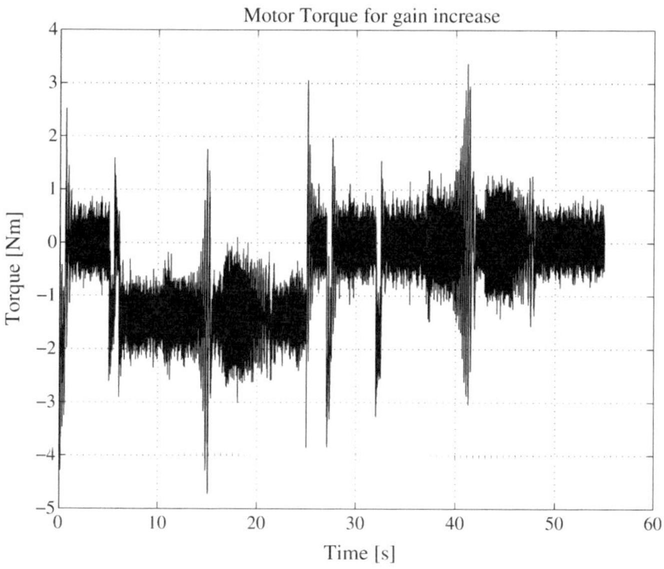

Fig. 8.7 Time response performance of the selected controller $\boldsymbol{\theta}_{DM}$ when gain is increased by a 2.5 factor

selected PID is capable of controlling the system. In Table 8.3 the maximum values achieved for each one of the indicators of the benchmark, for Set-1 and Set-2, are depicted. In all cases, imposed constraints for settling time and control effort are fulfilled.

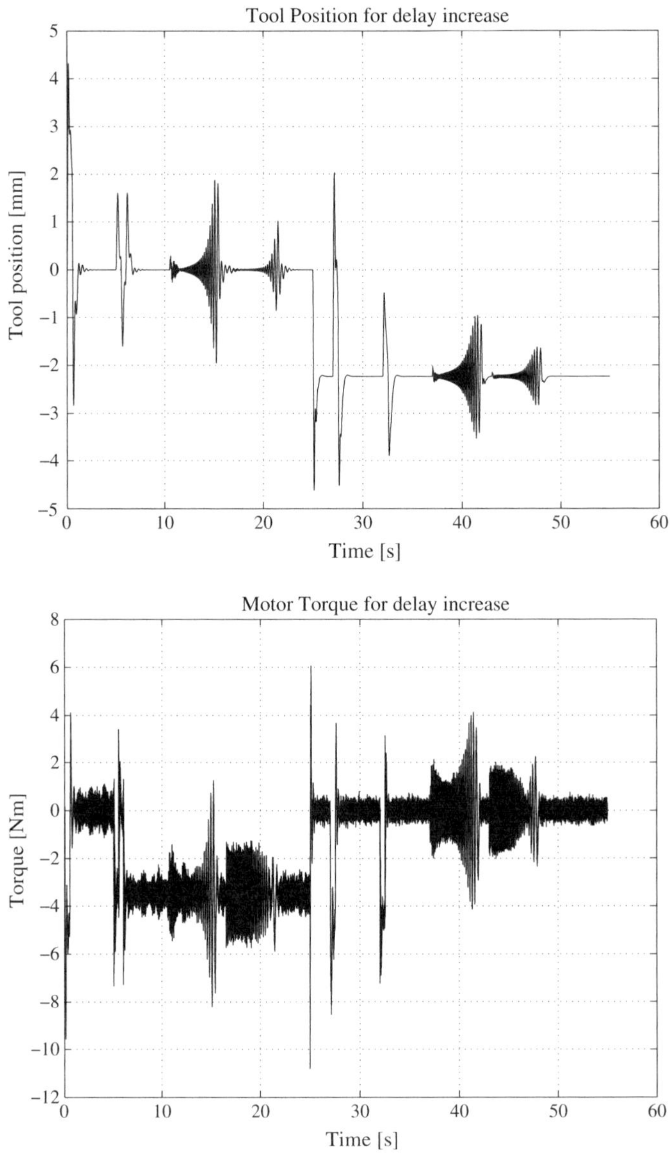

Fig. 8.8 Time response performance of the selected controller $\boldsymbol{\theta}_{DM}$ when delay is increased by a factor of 2

Table 8.3 Maximum values of time domain performance measures for model variations

Parameter (mm)	Set-1	Set-2
e_1	9.6842	11.5044
e_2	3.5487	3.6288
e_3	5.2386	5.4760
e_4	1.9683	1.7343
e_5	9.2335	11.2478
e_6	4.1597	4.4673
e_7	4.2118	4.7251
e_8	1.7679	1.8504
t_S^1	1.5265	1.0790
t_S^2	1.0585	0.5560
t_S^3	0.6955	0.6195
t_S^4	0.6799	0.5910
T_{NOISE}	1.0706	1.0639
T_{MAX}	10.8392	11.1216
T_{RMS}	1.4283	1.4482

Finally, the scores provided using the overall AOF defined or the benchmark are:

- $V_{Nom}(\boldsymbol{C}) = 62.6$
- $V_{Set-1}(\boldsymbol{C}) = 80.7$
- $V_{Set-2}(\boldsymbol{C}) = 81.9$
- $V(\boldsymbol{C}) = 142.9$

8.5 Conclusions

In this chapter, a digital controller was tuned in order to control the robotic arm of a manipulator. In this case, two sequential optimization instances were performed: the first one in order to get some knowledge on the trade-off expectations of the selected control structure; the second one (with knowledge retrieved from the first optimization and a redefined pertinency region) in order to achieve a more pertinent set of preferable solutions.

For this example, the MOOD procedure following a reliable based optimization was not possible, given the computational burden of simulating the model and a realistic implementation of the PID controllers (sampling rate and saturation). Due to this fact, robustness measures were used. In this case, it has been accepted that the AOF defined by the organizers was meaningful for the designer. If such AOF does not reflect its desired tradeoff, then simultaneous optimization using the same robustness indicators might be performed.

As a result, a suitable controller with an acceptable overall performance (using the full set of model uncertainties provided by the organizers) was achievable.

References

1. Moberg S, Ohr J, Gunnarsson S (2009) A benchmark problem for robust feedback control of a flexible manipulator. IEEE Trans Control Syst Technol 17(6):1398–1405
2. Reynoso-Meza G, Sanchis J, Blasco X, Martínez M (2010) Design of continuous controllers using a multiobjective differential evolution algorithm with spherical pruning. In: Applications of evolutionary computation. Springer, pp 532–541

Chapter 9
The 2012 IFAC Control Benchmark: An Industrial Boiler Process

Abstract In this chapter, a multiobjective optimization design procedure is applied to the multivariable version of the Boiler Control Problem, defined in 2nd IFAC Conference on Advances in PID Control, 2012. The chapter follows a *realistic* approach, closer to industrial practices: a nominal linear model will be identified and afterwards a constrained multiobjective problem with 5 design objectives will be stated. Such objectives will deal with overall robust stability, settling time performance and noise sensitivity. After approximating the Pareto Front and performing a multicriteria decision-making stage, the selected control system will be tested using the original nonlinear model.

9.1 Introduction

The process under consideration is the benchmark for PID control described in [5]. It proposes a boiler control problem [2, 4] based on the work of [7]. This benchmark version improves the model provided in [1] by adding a nonlinear combustion equation with a first order lag to model the excess oxygen in the stack and the stoichiometric *air-to-fuel* ratio for complete combustion. Several control proposals for the boiler can be found in [3, 6, 8, 10–12].

In order to propose a suitable controller for this benchmark, *quasi real* conditions will be followed, seeking to emulate the industrial tuning procedure that would be normally followed for such instances. *Quasi-real* conditions makes reference to the following steps:

1. Consider the (original) nonlinear model simulation as the *real process*.
2. Step tests are used to obtain simplified linear models from the *real process*.
3. Controllers are tuned using the aforementioned approximated models.
4. Selection procedure will be made according to experiments on the approximated models.
5. Selected controller will be implemented in the *real process*.

G. Reynoso Meza et al., *Controller Tuning with Evolutionary Multiobjective Optimization*, Intelligent Systems, Control and Automation: Science and Engineering 85, DOI 10.1007/978-3-319-41301-3_9

9.2 Benchmark Setup: Boiler Control Problem

The nonlinear explicit model is described by the following equations:

$$\dot{x_1}(t) = c_{11}x_4(t)x_1^{\frac{9}{8}} + c_{12}u_1(t-\tau_1) - c_{13}u_3(t-\tau_3) \tag{9.1}$$

$$\dot{x_2}(t) = c_{21}x_2(t) + \frac{c_{22}u_2(t-\tau_2) - c_{23}u_1(t-\tau_1) - c_{24}u_1(t-\tau_1)x_2(t)}{c_{25}u_2(t-\tau_2) - c_{26}u_1(t-\tau_1)} \tag{9.2}$$

$$\dot{x_3}(t) = -c_{31}x_1(t) - c_{32}x_4(t)x_1(t) + c_{33}u_3(t-\tau_3) \tag{9.3}$$

$$\dot{x_4}(t) = -c_{41}x_4(t) + c_{42}u_1(t-\tau_1) + c_{43} + n_5(t) \tag{9.4}$$

$$y_1(t) = c_{51}x_1(t-\tau_4) + n_1(t) \tag{9.5}$$

$$y_2(t) = c_{61}x_1(t-\tau_5) + n_2(t) \tag{9.6}$$

$$\begin{aligned} y_3(t) = {} & c_{70}x_1(t-\tau_6) + c_{71}x_3(t-\tau_6) \\ & + c_{72}x_4(t-\tau_6)x_1(t-\tau_6) + c_{73}u_3(t-\tau_3-\tau_6) \\ & + c_{74}u_1(t-\tau_1-\tau_6) \\ & + \frac{[c_{75}x_1(t-\tau_6) + c_{76}][1 - c_{77}x_3(t-\tau_6)]}{x_3(t-\tau_6)[x_1(t-\tau_6) + c_{78}]} \\ & + c_{79} + n_3(t) \end{aligned} \tag{9.7}$$

$$y_4(t) = [c_{81}x_4(t-\tau_7) + c_{82}]x_1(t-\tau_7) + n_4(t). \tag{9.8}$$

where $x_1(t), x_2(t), x_3(t), x_4(t)$ are the state variables of the system; $y_1(t)$, $y_2(t)$, $y_3(t)$, $y_4(t)$ the observed states; c_{ij}, τ_i and n_i are nonlinear coefficients, time constants and noise models, respectively, determined to improve the accuracy of the model. Finally, variables u_1, u_2 and u_3 are the inputs.

This benchmark version (Fig. 9.1) proposes a reduced 2×2 MIMO system with a measured load disturbance:

$$\begin{bmatrix} Y_1(s) \\ Y_3(s) \end{bmatrix} = \begin{bmatrix} P_{11}(s) & P_{13}(s) \\ P_{31}(s) & P_{33}(s) \end{bmatrix} \begin{bmatrix} U_1(s) \\ U_3(s) \end{bmatrix} + \begin{bmatrix} G_{1d}(s) \\ G_{3d}(s) \end{bmatrix} D(s) \tag{9.9}$$

where the inputs are fuel flow $U_1(s)$ [%] and water flow $U_3(s)$ [%], while the outputs are steam pressure $Y_1(s)$ [%] and water level $Y_3(s)$ [%]. $D(s)$ is a measured disturbance. This is a verified model, useful to propose, evaluate and compare different kinds of tuning/control techniques [3, 6, 9–11].

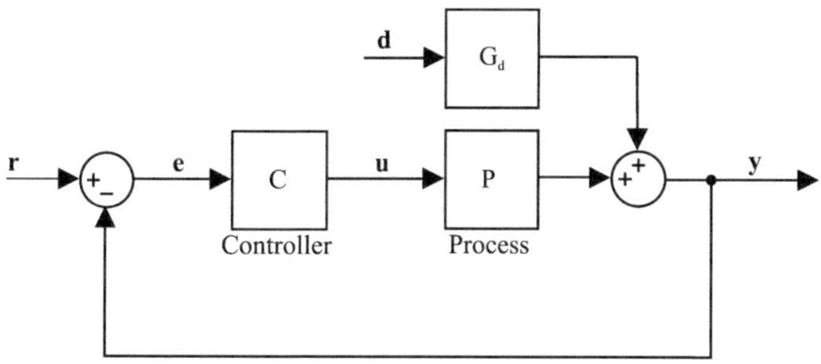

Fig. 9.1 Multivariable loop control

The proposed controller is:

$$C(s) = \begin{bmatrix} Kc_1\left(1 + \frac{1}{Ti_1 s}\right) & 0 \\ 0 & Kc_2\left(1 + \frac{1}{Ti_2 s}\right) \end{bmatrix}. \tag{9.10}$$

In [8], an identified linear model at the operating point is shown in Eqs. (9.11), (9.12) and depicted in Fig. 9.2.

$$P(s) = \begin{bmatrix} P_{11}(s) & P_{13}(s) \\ P_{31}(s) & P_{33}(s) \end{bmatrix} = \begin{bmatrix} \frac{0.3727e^{-3.1308s}}{55.68s+1} & \frac{-0.1642}{179.66s+1} \\ \frac{0.0055\cdot(166.95s-1)}{31.029s^2+s} & \frac{0.0106e^{-9.28s}}{s} \end{bmatrix}, \tag{9.11}$$

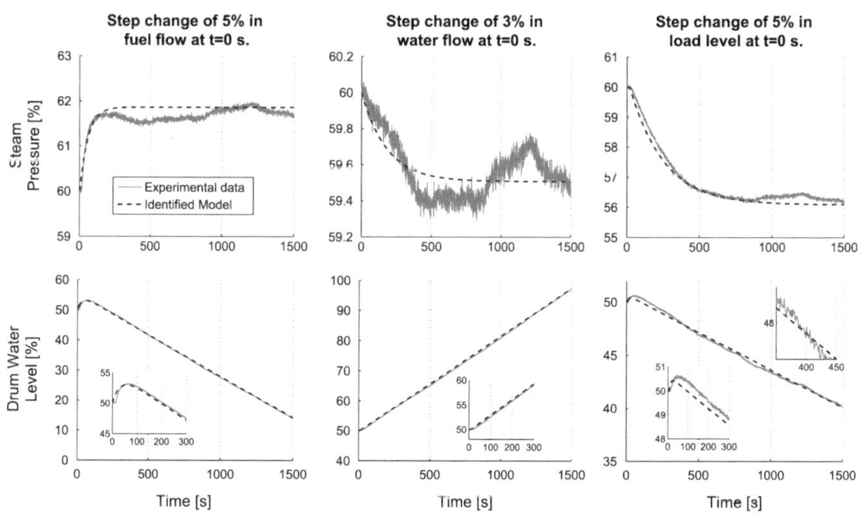

Fig. 9.2 Identified reduced model of the boiler process. Adapted from [8]

$$G_d(s) = \begin{bmatrix} G_{d1}(s) \\ G_{d3}(s) \end{bmatrix} = \begin{bmatrix} \frac{-0.78266e^{-17.841s}}{234.69s+1} \\ \frac{-0.0014079e^{-7.1872s}}{7.9091s^2+s} \end{bmatrix}. \tag{9.12}$$

9.3 The MOOD Approach

To deal with the boiler control problem, five objectives are defined:

$J_1(\boldsymbol{\theta})$: Settling time for $Y_1(s)$ at presence of a step load disturbance $D(s)$.

$$J_1(\boldsymbol{\theta}) = J_{t_{98\%}}(\boldsymbol{\theta}). \tag{9.13}$$

$J_2(\boldsymbol{\theta})$: Settling time for $Y_2(s)$ at presence of a step load disturbance $D(s)$.

$$J_1(\boldsymbol{\theta}) = J_{t_{98\%}}(\boldsymbol{\theta}). \tag{9.14}$$

$J_3(\boldsymbol{\theta})$: Biggest log modulus (BLT) for overall robustness.

$$J_3(\boldsymbol{\theta}) = 20 \log \left| \frac{W(s)}{1 + W(s)} \right|, \tag{9.15}$$
$$W(s) = -1 + \det\left(I + P(s)C_{\boldsymbol{\theta}}(s)\right).$$

$J_4(\boldsymbol{\theta})$: Maximum value of noise sensitivity function M_u for loop 1 (Eq. 2.6).

$$J_4(\boldsymbol{\theta}) = \left\| C_{1,\boldsymbol{\theta}}(s)(1 + P_{11}(s)C_{1,\boldsymbol{\theta}}(s))^{-1} \right\|_{\infty}. \tag{9.16}$$

$J_5(\boldsymbol{\theta})$: Maximum value of noise sensitivity function M_u for loop 2 (Eq. 2.6).

$$J_5(\boldsymbol{\theta}) = \left\| C_{2,\boldsymbol{\theta}}(s)(1 + P_{22}(s)C_{2,\boldsymbol{\theta}}(s))^{-1} \right\|_{\infty}. \tag{9.17}$$

Therefore the MOP to solve will be:

$$\min_{\boldsymbol{\theta}} \boldsymbol{J}(\boldsymbol{\theta}) = [J_1(\boldsymbol{\theta}), J_2(\boldsymbol{\theta}), J_3(\boldsymbol{\theta}), J_4(\boldsymbol{\theta}), J_5(\boldsymbol{\theta})] \tag{9.18}$$
$$\boldsymbol{\theta} = [Kc_1, Ti_1, Kc_2, Ti_2] \tag{9.19}$$

subject to:

$$0 \leq Kc_{1,2} \leq 1 \tag{9.20}$$
$$0 < Ti_{1,2} \leq 1 \tag{9.21}$$
$$\text{Stable in closed loop.} \tag{9.22}$$

As five design objectives are stated, the sp-MODE-II algorithm with parameters shown in Table 9.2 will be used (details in Chap. 3). Preference matrix $\mathfrak{P}$ is defined in Table 9.1. In this case, just three objectives will be used in the MCDM phase ($J_1(\boldsymbol{\theta})$, $J_2(\boldsymbol{\theta})$ and $J_3(\boldsymbol{\theta})$).

After the optimization process, Pareto Front and Set approximations $\boldsymbol{J}^*_P$ and $\boldsymbol{\Theta}^*_P$ are calculated. It is important to remark that, there are some *apparently* dominated solutions in the 3D plot (Fig. 9.3a), however in the original 5-dimensional space they are non-dominated. Performing DM phase with just three design objectives is due to its preferability according to the preference matrix stated in Table 9.1. In Fig. 9.4 additional information regarding time responses with the test used for optimization are depicted.

Table 9.1 Preference matrix for GPP index. Five preference ranges have been defined: *Highly Desirable* (HD), *Desirable* (D), *Tolerable* (T) *Undesirable* (U) and *Highly Undesirable* (HU)

Preference matrix $\mathfrak{P}$						
Objective	← HD →	← D →	← T →	← U →	← HU →	
	J_q^0	J_q^1	J_q^2	J_q^3	J_q^4	J_q^5
$J_1(\boldsymbol{\theta})$ (s)	300	400	600	800	1500	2000
$J_2(\boldsymbol{\theta})$ (s)	600	800	1000	1500	1800	2000
$J_3(\boldsymbol{\theta})$ (–)	0	1	4	6	8	16
$J_4(\boldsymbol{\theta})$ (dB)	0.0	5.0	8.0	10	20	25
$J_5(\boldsymbol{\theta})$ (dB)	0.0	5.0	8.0	10	20	25

Table 9.2 Parameters used for sp-MODE-II. Further details in Chap. 3

Scope	Parameter	Value
	Parameter	**Value**
sp-MODE-II / sp-MODE / DE	**Evolutionary mechanism**	
sp-MODE-II / sp-MODE / DE	F (Scaling factor)	0.5
sp-MODE-II / sp-MODE / DE	Cr (Crossover rate)	0.9
sp-MODE-II / sp-MODE / DE	N_p (Population)	50
sp-MODE-II / sp-MODE	**Pruning mechanism**	
sp-MODE-II / sp-MODE	$\boldsymbol{\beta}^{\epsilon}$ (Arcs)	30
sp-MODE-II	**Pertinency mechanism**	
sp-MODE-II	J_{GPP}^{max}	$J_{GPP}([800, 1500, 6, 10, 10])$
sp-MODE-II	$\overline{car(J_P^*)}$	30

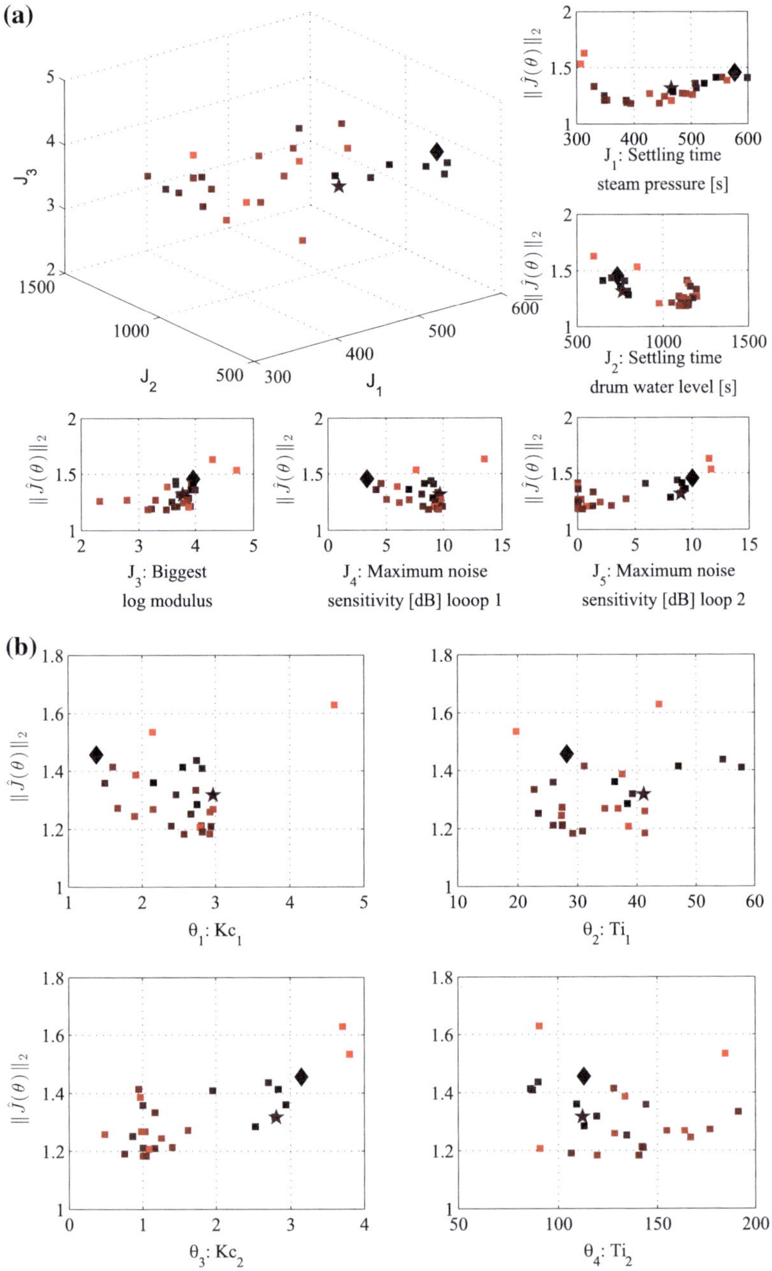

Fig. 9.3 Pareto front and pareto set approximated for the boiler problem. Colored according to GPP index, *darker color* corresponds to lower GPP index. ♦ solution with the lowest GPP index, ★ the DM's choice. **a** Pareto Front approximation $\boldsymbol{J}_p^*$. **b** Pareto Set approximation $\boldsymbol{\Theta}_p^*$

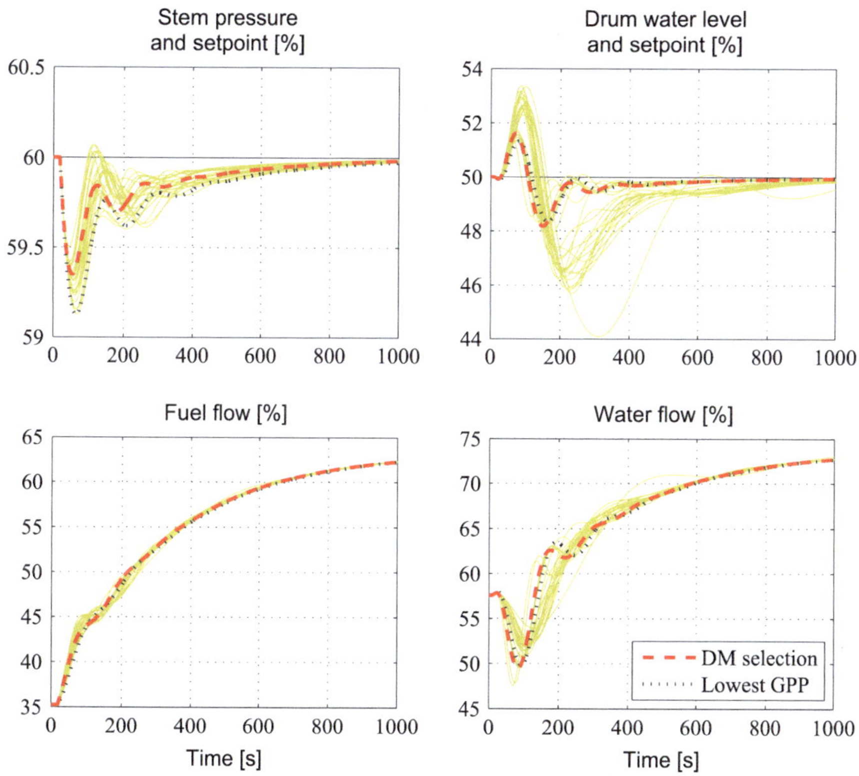

Fig. 9.4 Time responses of the approximated pareto set for the boiler benchmark

In the MCDM stage, tradeoff among solutions is compared and analysed. Solution with lowest GPP value is depicted with ⧫ while the DM's choice with a ★. The latter solution has been preferred over the former due to its improvement on settling time for steam pressure, in exchange of noise sensitivity in the same loop. Remember that sp-MODE-II approach enables us to approximate a pertinent and compact Pareto Front approximation $\boldsymbol{\Theta}_P^*$ around the preferable region, according to the preference matrix $\mathfrak{P}$. Selected solution $\boldsymbol{\theta}_{DM} = [2.9672, 41.1272, 2.8046, 112.0901]$ leads to the following multivariable controller, which will undergo further control test evaluations.

$$C_{DM}(s) = \begin{bmatrix} 2.9672\left(1 + \frac{1}{41.1272s}\right) & 0 \\ 0 & 2.8046\left(1 + \frac{1}{112.0901s}\right) \end{bmatrix}. \tag{9.23}$$

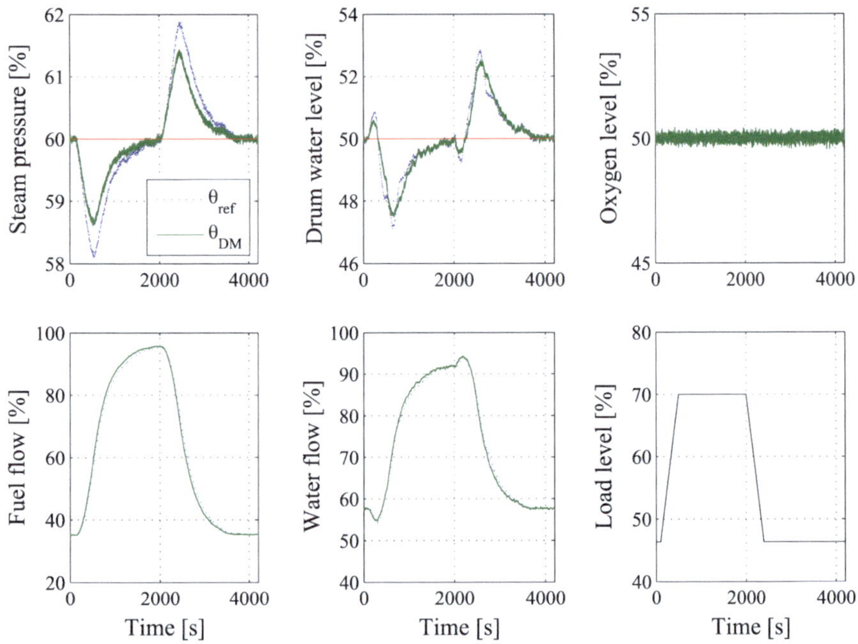

Fig. 9.5 PI controller θ_{DM} compared with the reference case θ_{ref} for Test-1

9.4 Control Tests

In this section the selected solution $\boldsymbol{\theta}_{DM}$ will be tried out with the tests proposed in the original benchmark. The two different tests are:

Test-1: Performance when the system has to attend a time variant load level.
Test-2: Performance when the system has to attend a sudden change in the steam pressure set-point.

In order to evaluate the overall performance of a given controller, the benchmark defined an index $I_{benchmark}(C_e, C_{ref}, \omega)$ which is automatically calculated when running a test on the benchmark (further details are available in [5]). Such index is an aggregate objective function, which combines ratios of the IAE Eq. (2.9), ITAE equation (2.10) and TV equation[1] (2.18). These ratios are calculated as the relations between the proposal to evaluate $C_e = \boldsymbol{\theta}_{DM}$ and a reference controller C_{ref}. The aggregation uses a weighting factor ω for the ratios of the control action values ($\omega \dot{T} V$). In the original benchmark, two PI controllers $\boldsymbol{\theta}_{ref} = [2.5, 50, 1.25, 50]$ are used as C_{ref} and the weighting factor is set to $\omega = 0.25$.

Figures 9.5 and 9.6 compares closed loop results of the selected controller $\boldsymbol{\theta}_{DM}$ with reference controller $\boldsymbol{\theta}_{ref}$ for both tests. In Test-1, $\boldsymbol{\theta}_{DM}$ controller has a better

[1]also known as IADU.

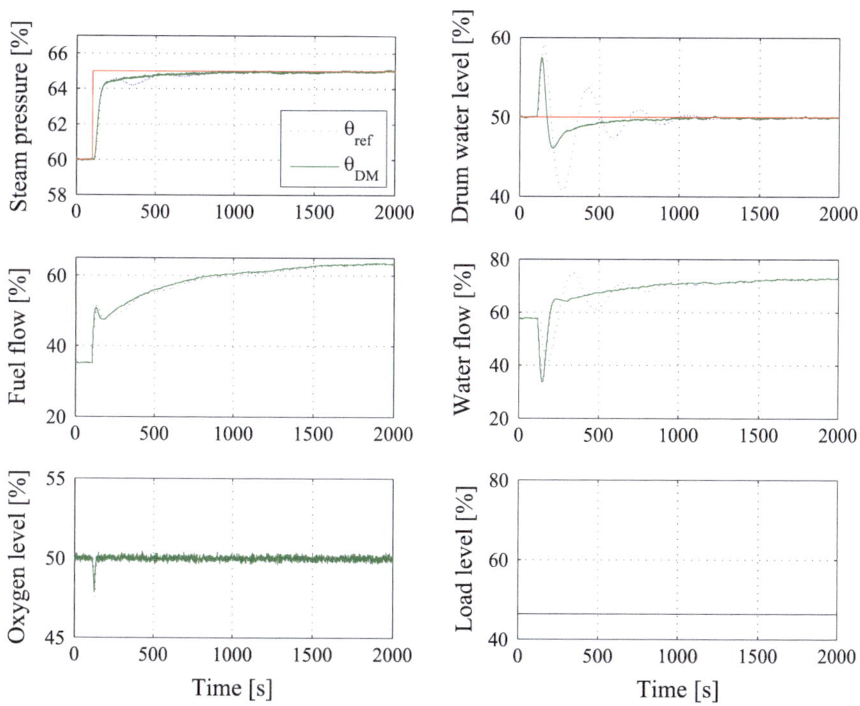

Fig. 9.6 PI controller θ_{DM} compared with the reference case θ_{ref} for Test-2

Table 9.3 $I_{benchmark}(\boldsymbol{\theta}_{DM}, \boldsymbol{\theta}_{ref}, 0.25)$ performance achieved of the selected design alternative for Test-1 and Test-2

	$I_{benchmark}(\boldsymbol{\theta}_{DM}, \boldsymbol{\theta}_{ref}, 0.25)$
Test-1	0.9546
Test-2	0.7993

response for the steam pressure loop, by minimizing the disturbance effect by the time variant load level. In Test-2, $\boldsymbol{\theta}_{DM}$ controller achieves a smother response compared with $\boldsymbol{\theta}_{ref}$ in the drum water level control loop. For comparison purposes, the benchmark index $I_{benchmark}(\boldsymbol{\theta}_{DM}, \boldsymbol{\theta}_{ref}, 0.25)$ has been calculated for both tests (Table 9.3). In both cases, as $I_{benchmark}(\boldsymbol{\theta}_{DM}, \boldsymbol{\theta}_{ref}, 0.25)$ is below 1, controller $\boldsymbol{\theta}_{DM}$ has a better performance than $\boldsymbol{\theta}_{r}$, concluding that (in terms of the preferability of the benchmark organizers) the additional control effort provided by the $\boldsymbol{\theta}_{DM}$ compensates the performance improvement in the remainder indicators. Therefore, selected controller $\boldsymbol{\theta}_{DM}$ provides an improvement on the overall MIMO loop performance.

9.5 Conclusions

In this chapter, a multivariable controller was tuned using a MOOD procedure. In this case, the MOOD procedure followed a *realistic* approach, closer to industrial practices as:

1. Experiments were performed in order to obtain a simplified linear model.
2. Controllers were adjusted with the approximated model.
3. The decision making process was made according to performance on the approximated models.
4. The selected controller was finally implemented in the nonlinear process.

As a result, a compact and pertinent Pareto Front was approximated. After the MCDM stage, a suitable controller was selected, showing a better performance, when compared with the reference controller of the benchmark.

References

1. Bell R, Åström KJ (1987) Dynamic models for boiler-turbine alternator units: Data logs and parameter estimation for a 160 MW unit. Technical Report ISRN LUTFD2/TFRT–3192–SE, Department of Automatic Control, Lund University, Sweden
2. Fernández I, Rodríguez C, Guzmán J, Berenguel M (2011) Control predictivo por desacoplo con compensación de perturbaciones para el benchmark de control 2009–2010. Revista Iberoamericana de Automática e Informática Industrial Apr. 8(2):112–121
3. Garrido J, Márquez F, Morilla F (2012) Multivariable PID control by inverted decoupling: application to the benchmark PID 2012. In: Proceedings of the IFAC conference on advances in PID control (PID'12), March 2012
4. Morilla F (2010) Benchmark 2009-10 grupo temático de ingeniera de control de CEA-IFAC: Control de una caldera. Febrero 2010. http://www.dia.uned.es/~fmorilla/benchmark09_10/
5. Morilla, F. Benchmar for PID control based on the boiler control problem. http://servidor.dia.uned.es/~fmorilla/benchmarkPID2012/, 2012. Internal report, UNED Spain
6. Ochi Y (2012) PID controller design for MIMO systems by applying balanced truncation to integral-type optimal servomechanism. In: Proceedings of the IFAC conference on advances in PID Control (PID'12), March 2012
7. Pellegrinetti G, Bentsman J (1996) Nonlinear control oriented boiler modeling-a benchmark problem for controller design. IEEE Trans Control Syst Technol 4(1):57–64
8. Reynoso-Meza G, Sanchis J, Blasco X, Martínez MA (2016) Preference driven multi-objective optimization design procedure for industrial controller tuning. Inf Sci 339:105–131
9. Rojas JD, Morilla F, Vilanova R (2012) Multivariable PI control for a boiler plant benchmark using the virtual reference feedback tuning. In: Proceedings of the IFAC conference on advances in PID control (PID'12), March 2012
10. Saeki M, Ogawa K, Wada N (2012) Application of data-driven loop-shaping method to multi-loop control design of benchmark PID 2012. In: Proceedings of the IFAC conference on advances in PID control (PID'12), March 2012

11. Silveira A, Coelho A, Gomes F (2012) Model-free adaptive PID controllers applied to the benchmark PID12. In: Proceedings of the IFAC conference on advances in PID control (PID'12), March 2012
12. Schez HS, Reynoso-Meza G, Vilanova R, Blasco X (2015) Multistage procedure for pi controller design of the boiler benchmark problem. In: 2015 IEEE 20th conference on emerging technologies factory automation (ETFA), Sept 2015, pp 1–4

Part IV
Applications

This part is dedicated to implement the multiobjective optimization design (MOOD) procedure for controller tuning in real processes. The aim of this part is to link this procedure with the implementation phase; therefore, emphasis on the decision making stage is given. Also, general guidelines to close the gap between optimization and decision making are provided in each case.

Chapter 10
Multiobjective Optimization Design Procedure for Controller Tuning of a Peltier Cell Process

Abstract In this chapter a Peltier cell is used for cooling and freezing purposes. The main challenge from the control point of view is to guarantee the setpoint response performance for both tasks despite the process nonlinearities. For this purpose, reliability based optimization approach is stated and tackled with the multiobjective optimization design procedure.

10.1 Introduction

A *Reliability Based Design Optimization* (RBDO) instance [2] might be useful in order to anticipate any degradation on control performance due to unexpected or unmodelled process dynamics. It might be also useful in processes where their dynamics change between operational areas due to nonlinearities. For example, in [7] such approach was used in a Peltier process, where different dynamics are expected when the device is working around its *cool* or its *freeze* operational area.

Peltier cells are basically thermoelectric heat pumps, which use two different semiconductors connected electrically in series and thermally in parallel. These semiconductors are sandwiched between two ceramic plates in order to create a heat flow from one plate (cold-face) to the other (hot-face). Peltier cells can be found in several applications as thermoelectric generators [1], hypothermic treatment [5], cooling systems in photovoltaic cells [6] and laser ablation cells [3].

In this chapter, a multiobjective RBDO instance is proposed, in order to tune a PI controller which controls the *cooling* and *freezing* dynamics of a Peltier cell. Since different linear models will be identified around both operational regions, the performance degradation of the controller will be evaluated in order achieving a reliable measure of the controller performance.

G. Reynoso Meza et al., *Controller Tuning with Evolutionary Multiobjective Optimization*, Intelligent Systems, Control and Automation: Science and Engineering 85, DOI 10.1007/978-3-319-41301-3_10

10.2 Process Description

A Peltier cell (Fig. 10.1) is a device based on the *Peltier scheme*. It is a heat pump where the manipulated variable u is the voltage (in [%] of its range) applied to the cell and the controlled variable is the temperature [°C] of the cold-face T_{cold}. Peltier effect is modeled as follows [4]:

$$\dot{Q} = \alpha \cdot T_{cold} \cdot I, \tag{10.1}$$

where $\dot{Q}$ is the heat power, T_{cold} the temperature, I the current and α is known as the *Seebeck* coefficient.

This kind of processes have nonlinear dynamics due to the Peltier effect.

The main goal of the control loop (Fig. 10.2) is to keep the desired temperature within the operational range $T_{op} = [-12.0,\ 6.0]$ °C comprising the *cool* region ($\approx$ 4.0 °C) and the *freeze* one ($\approx$ −8.0 °C). The desired performance should be achieved in the whole operational range despite the nonlinear dynamics due to the Peltier effect.

Before going further into the MOOD procedure, a model will be identified. Thus, temperature responses to consecutive input changes within T_{op} interval are measured.

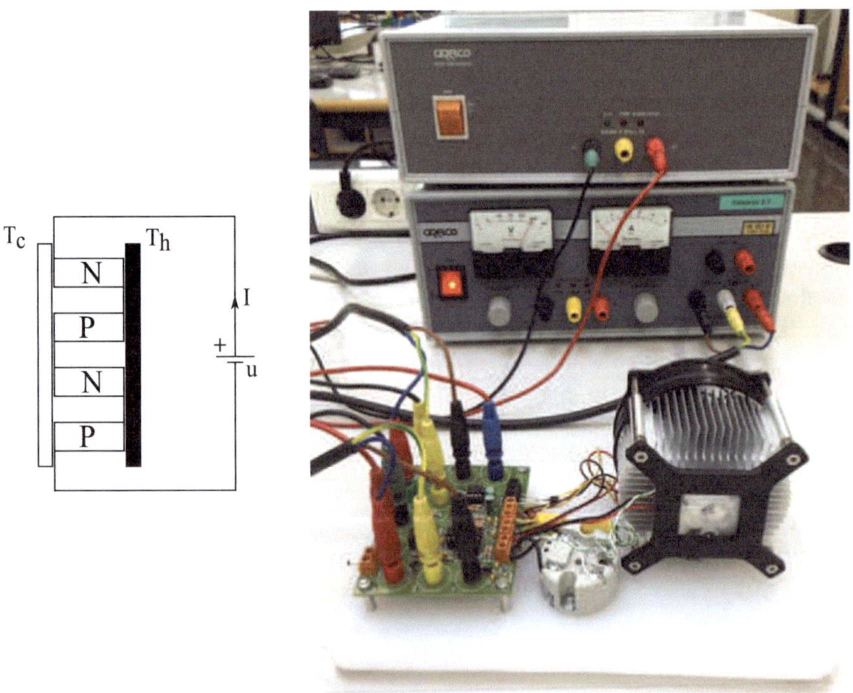

Fig. 10.1 Peltier cell sketch (*left*). Peltier cell laboratory set-up (*right*)

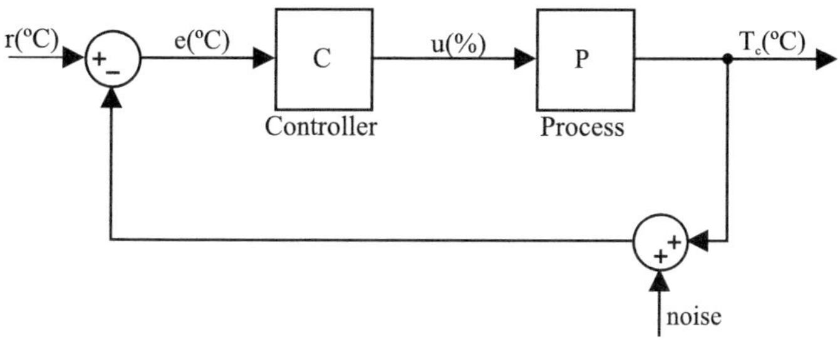

Fig. 10.2 Basic loop for PI control

As a result several first order plus dead time models (FOPDT) are identified:

$$P(s) = \frac{K}{\tau\, s + 1} e^{-Ls},$$

where K [°C/%] is the process gain, τ [s] the time constant and L [s] system delay. Figure 10.3a, b depict temperature responses for the *cool* and *freeze* zones respectively. The resulting models are shown in Tables 10.1 and 10.2. Identification was performed with the identification toolbox available in Matlab©. Notice the difference between models concerning K and τ values, which agree with the non-linearity nature of the system.

10.3 The MOOD Approach

The challenge is using just one controller to control both zones (Fig. 10.2). Nominal models selected for *cooler* $P_C(s)$ and *freezer* $P_f(s)$ are:

$$P_C(s) = \frac{-0.6030}{3.3166s + 1} e^{-0.2s}, \tag{10.2}$$

$$P_F(s) = \frac{-0.3155}{3.1921s + 1} e^{-0.4s}, \tag{10.3}$$

where $P_C(s)$ includes a delay of $L = 0.2$ which is the control period and $P_F(s)$ has a value of $L = 0.4$. Since the controller must be able to manage both operational zones dealing with the nonlinearities, two different sets of FOPDT models (Φ_C and Φ_F) are defined around each nominal model. These sets contain 51 models, randomly sampled from the intervals $K = -0.6030 \pm 50$ %, $\tau = 3.3166 \pm 30$ %, $L = 0.2$ for Φ_C and $K = -0.3155 \pm 50$ %, $\tau = 3.1921 \pm 30$ %, $L = 0.4 \pm 0.2$ for Φ_F.

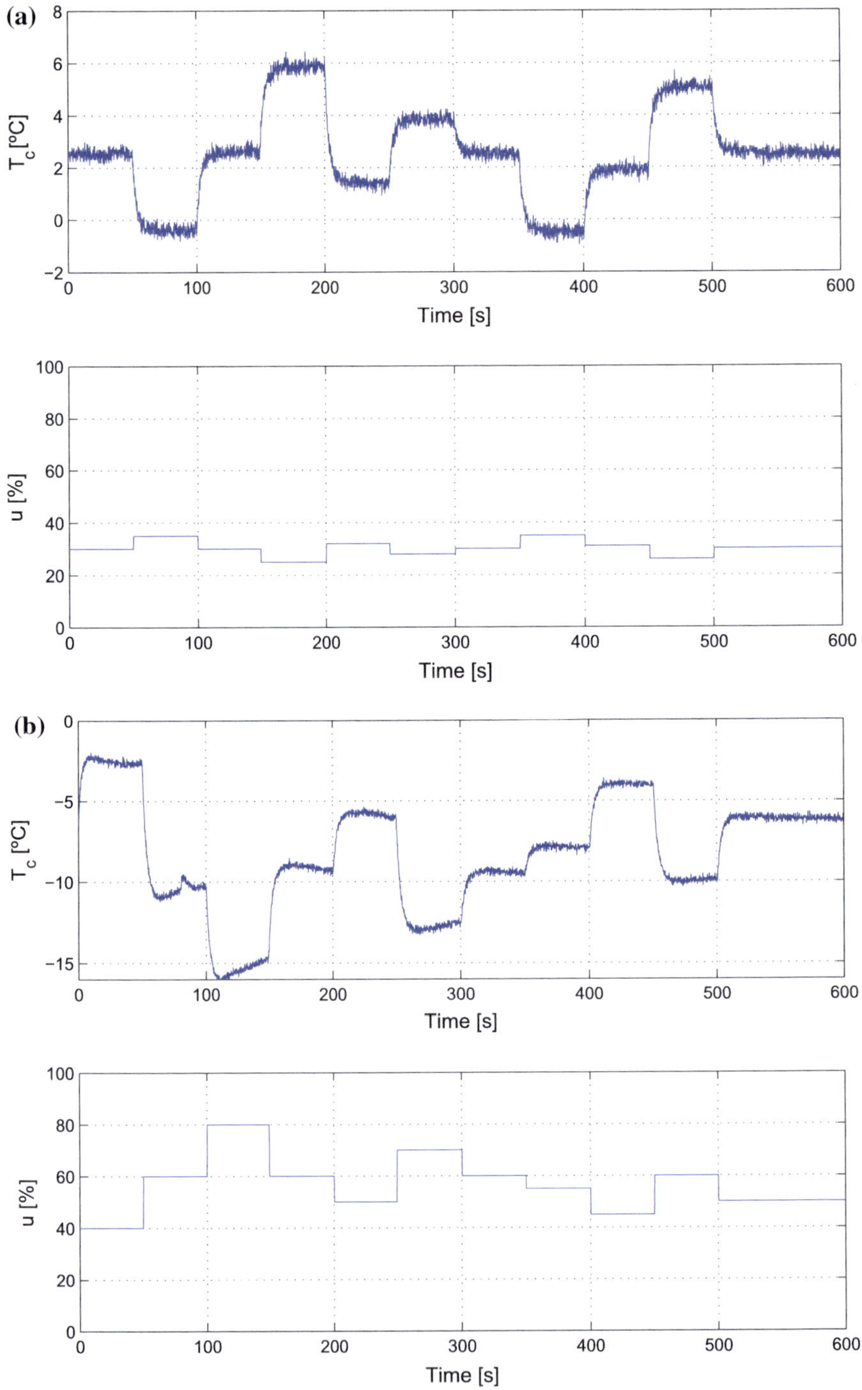

Fig. 10.3 Experiments for identification. **a** Cooler. **b** Freezer

Table 10.1 Identified models for cool region

Input change (%)	K (°C/%)	L (s)	τ (s)
30 to 35	−0.6162	0.0	2.9887
35 to 30	−0.5780	0.0	3.0782
30 to 25	−0.6140	0.0	3.2742
25 to 32	−0.6839	0.0	3.0954
28 to 30	−0.7684	0.0	2.5281
30 to 35	−0.6030	0.0	3.3166
35 to 31	−0.6224	0.0	3.1063
31 to 26	−0.5716	0.0	3.4106

Table 10.2 Identified models for freeze region

Input change (%)	K (°C/%)	L (s)	τ (s)
40 to 60	−0.3948	0.24	2.4514
60 to 80	−0.2817	0.08	2.7599
80 to 60	−0.3640	0.14	2.7887
60 to 50	−0.3538	0.16	2.9167
50 to 70	−0.3155	0.02	3.1921
70 to 60	−0.2786	0.62	2.7788
60 to 55	−0.3808	0.23	3.1359
55 to 45	−0.3913	0.20	3.2456
45 to 60	−0.4087	0.00	3.0051

The controller to tune being

$$C_{\boldsymbol{\theta}}(s) = K_c \cdot \left(1 + \frac{1}{T_i s}\right)$$

and the decision variables

$$\boldsymbol{\theta} = [K_c, T_i],$$

the following MOP statement is defined:

$$\min_{\boldsymbol{\theta}} \boldsymbol{J}(\boldsymbol{\theta}) = [J_1(\boldsymbol{\theta}), \cdots, J_7(\boldsymbol{\theta})] \tag{10.4}$$

where design objectives are:

$J_1(\boldsymbol{\theta})$: Median settling time for a setpoint step change within the *cool* operational zone using set Φ_C.

$$J_1(\boldsymbol{\theta}) = \varsigma_{median} = median(\boldsymbol{\varsigma}) \\ \varsigma_i = J_{t_{98\,\%}}(\boldsymbol{\theta}, \phi_i), \forall \phi_i \in \Phi_C \tag{10.5}$$

$J_2(\boldsymbol{\theta})$: Maximum value of the sensitivity function for the *cooler* nominal loop.

$$J_2(\boldsymbol{\theta}) = \left\| (1 + P_C(s)C_{\boldsymbol{\theta}}(s))^{-1} \right\|_\infty \tag{10.6}$$

$J_3(\boldsymbol{\theta})$: Median settling time for a setpoint step change within the *freeze* operational zone using set Φ_F.

$$J_3(\boldsymbol{\theta}) = \varsigma_{median} = median(\boldsymbol{\varsigma}) \\ \varsigma_i = J_{t_{98\,\%}}(\boldsymbol{\theta}, \phi_i), \forall \phi_i \in \Phi_F \tag{10.7}$$

$J_4(\boldsymbol{\theta})$: Maximum value of the sensitivity function for the freezer nominal loop.

$$J_4(\boldsymbol{\theta}) = \left\| (1 + P_F(s)C_{\boldsymbol{\theta}}(s))^{-1} \right\|_\infty \tag{10.8}$$

$J_5(\boldsymbol{\theta})$: Median of the differences between settling time with a different model of the set Φ_C and $J_1(\boldsymbol{\theta})$ of the cooler nominal model.

$$J_5(\boldsymbol{\theta}) = median(\boldsymbol{\varsigma}) \\ \varsigma_i = |J_{t_{98\,\%}}(\boldsymbol{\theta}, \phi_i) - J_1(\boldsymbol{\theta})|, \forall \phi_i \in \Phi_C \tag{10.9}$$

$J_6(\boldsymbol{\theta})$: Median of the differences between settling time with a different model of the set Φ_F and $J_3(\boldsymbol{\theta})$ of the freezer nominal model.

$$J_6(\boldsymbol{\theta}) = median(\boldsymbol{\varsigma}) \\ \varsigma_i = |J_{t_{98\,\%}}(\boldsymbol{\theta}, \phi_i) - J_3(\boldsymbol{\theta})|, \forall \phi_i \in \Phi_F \tag{10.10}$$

$J_7(\boldsymbol{\theta})$: High frequency gain of the controller (noise sensitivity).

$$J_7(\boldsymbol{\theta}) = \|C_{\boldsymbol{\theta}}(j\omega)\|_{\omega\in[1\ldots10^3]}. \tag{10.11}$$

Design objectives $J_1(\boldsymbol{\theta})$, $J_3(\boldsymbol{\theta})$, $J_5(\boldsymbol{\theta})$ and $J_6(\boldsymbol{\theta})$ are performance objectives based on reliability-based measures (they are calculated using a set of models). Design objectives $J_2(\boldsymbol{\theta})$, $J_4(\boldsymbol{\theta})$ are design objectives used as robustness measures based on nominal models. Finally, design objective $J_7(\boldsymbol{\theta})$ gives a measure of noise sensibility; as it can be appreciated in Fig. 10.4, the measured signal T_{cold} presents random noise in the range ±0.225 °C (more or less four times the quantization error of the analog to digital converter).

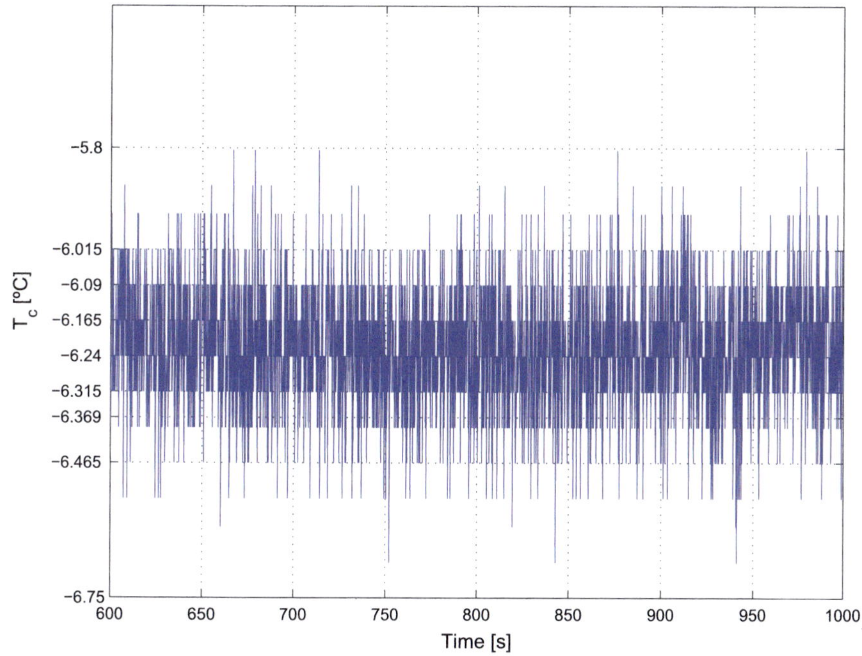

Fig. 10.4 Noisy temperature measures T_{cold} ($u = 50\,\%$), freeze region

Therefore, the MOP statement is:

$$\min_{\theta} \boldsymbol{J}(\boldsymbol{\theta}) = [J_1(\boldsymbol{\theta}), \cdots, J_7(\boldsymbol{\theta})] \tag{10.12}$$

$$\boldsymbol{\theta} = [K_c, T_i] \tag{10.13}$$

subject to:

$$0 \leq K_c \leq 10 \tag{10.14}$$

$$0 \leq T_i \leq 1000 \tag{10.15}$$

$$\text{Stable in closed loop.} \tag{10.16}$$

To deal with many objectives in the EMO phase the sp-MODE-II algorithm [8] will be used with the preference matrix shown in Table 10.3 and algorithm's parameters of Table 10.4.

Table 10.3 Preference matrix $\mathfrak{P}$ for GPP index with five preference ranges: *Highly Desirable* (HD), *Desirable* (D), *Tolerable* (T) *Undesirable* (U) and *Highly Undesirable* (HU)

Preference matrix $\mathfrak{P}$						
Objective	←HD→	← D→	← T→	← U→	← HU→	
	J_q^0	J_q^1	J_q^2	J_q^3	J_q^4	J_q^5
$J_1(\boldsymbol{\theta})$ (s)	0.0	5.0	10.0	15.0	20.0	30.0
$J_2(\boldsymbol{\theta})$ (-)	1.0	1.4	1.5	1.6	1.8	2.0
$J_3(\boldsymbol{\theta})$ (s)	0.0	10.0	20.0	25.0	20.0	30.0
$J_4(\boldsymbol{\theta})$ (-)	1.0	1.4	1.5	1.6	1.8	2.0
$J_5(\boldsymbol{\theta})$ (s)	0.0	0.5	1.0	2.0	10.0	20.0
$J_6(\boldsymbol{\theta})$ (s)	0.0	0.5	1.0	2.0	10.0	20.0
$J_7(\boldsymbol{\theta})$ (dB)	0.0	1.0	5.0	10.0	40.0	45.0

Table 10.4 Parameters used for sp-MODE-II. (Further details in Chap. 3)

Algorithm	Parameter	Value
	Evolutionary mechanism	
sp-MODE-II / sp-MODE / DE	F (Scaling factor)	0.5
sp-MODE-II / sp-MODE / DE	Cr (Crossover rate)	0.9
sp-MODE-II / sp-MODE / DE	N_p (Population)	20
	Pruning mechanism	
sp-MODE-II / sp-MODE	$\boldsymbol{\beta}_\epsilon$ (Arcs)	40
	Pertinency mechanism	
sp-MODE-II	J_{GPP}^{max}	$J_{GPP}([50, 15, 10, 1.8, 50])$
sp-MODE-II	$\overline{car(J_P^*)}$	40

In Fig. 10.5 the Pareto Front and Set approximations obtained are shown. After an analysis of such approximations, the MCDM phase returns three controllers, selected for further control tests:

$$C_{1_{DM}}(s) = 2.45 \cdot \left(1 + \frac{1}{1.27s}\right), \tag{10.17}$$

$$C_{2_{DM}}(s) = 2.01 \cdot \left(1 + \frac{1}{2.60s}\right), \tag{10.18}$$

$$C_{3_{DM}}(s) = 0.86 \cdot \left(1 + \frac{1}{0.89s}\right). \tag{10.19}$$

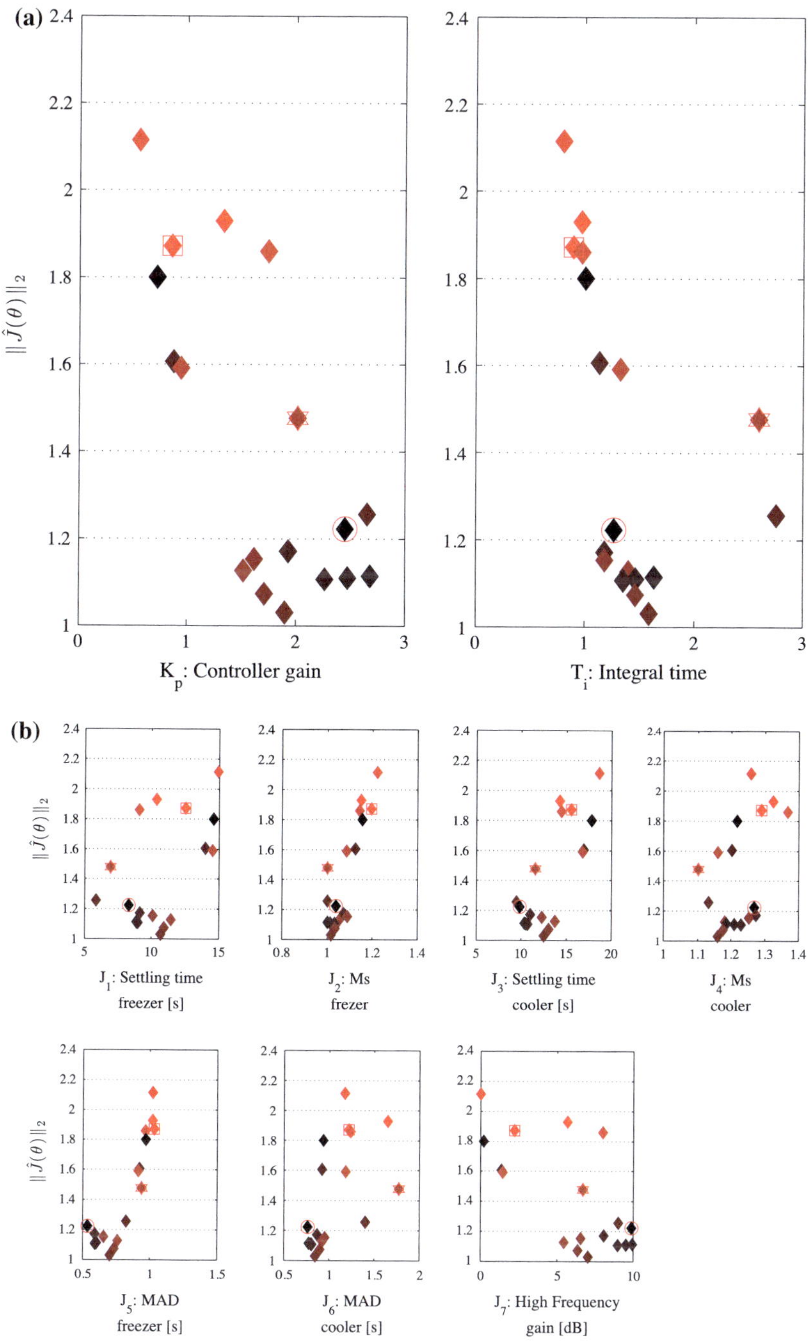

Fig. 10.5 Pareto set and pareto front approximations and selected controllers $C_{1_{DM}}$ (*square*), $C_{2_{DM}}$ (*star*) and $C_{3_{DM}}$ (*circle*). **a** Pareto Set. **b** Pareto Front

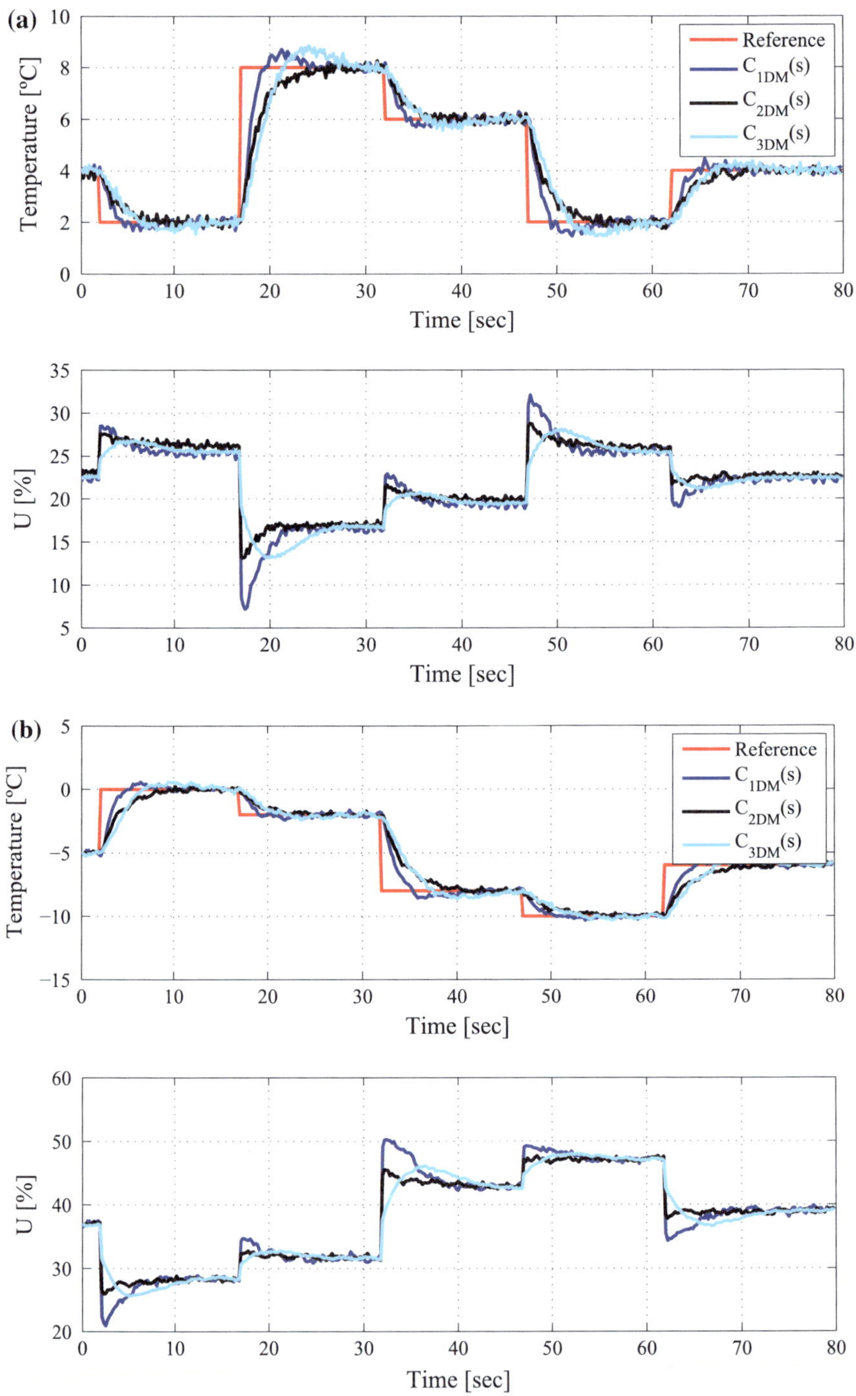

Fig. 10.6 Control tests on selected controllers. **a** Cooler. **b** Freezer

Such controllers have different noise sensibilities ($J_7(\boldsymbol{\theta})$): $C_{1_{DM}}$ has the worst sensibility (within the approximated Pareto Front), while $C_{3_{DM}}$ has the better. As commented before, noisy measurements oscillate around $T_{cold} \pm 0.225$ °C. Then, considering this effect on the MCDM in order to select a subset of feasible controllers for a further analysis is a reasonable decision.

10.4 Control Tests

Selected controllers $C_{1_{DM}}(s)$, $C_{2_{DM}}(s)$ and $C_{3_{DM}}(s)$ will undergo various control real tests on the Peltier device. This final step in the decision making process in order to select the most preferable controller, is necessary since we have no idea how it is going to affect noise issue to time performances of the controllers. Therefore, this final analysis with a small subset of selected solutions is a natural process, in order to verify real performances in the real platform.

Performance evaluation for cool and freeze zones with different setpoint responses are depicted in Fig. 10.6. Additional indicators are provided in Tables 10.5, 10.6, 10.7 and 10.8.

In Tables 10.5 and 10.6 closed loop settling time responses are shown. Also, obtained values of $J_1(\boldsymbol{\theta})$, $J_5(\boldsymbol{\theta})$ (freeze region) and $J_2(\boldsymbol{\theta})$, $J_6(\boldsymbol{\theta})$ (cool region) are indicated. Notice that $C_{3_{DM}}(s)$ is, in general, closer to the predicted performances in the optimization process (median value and median deviation) however $C_{1_{DM}}(s)$ and $C_{2_{DM}}(s)$ are not close to the expected values. It is due to the noise and quantization effects, not considered a priori in the optimization process. But objective $J_7(\boldsymbol{\theta})$ was included to appreciate the implication of having such performance in the nominal models when compared with the high frequency gain of the controller. It means controllers with better $J_7(\boldsymbol{\theta})$ are more likely to have the expected performances when they are controlling the real process.

In Tables 10.7 and 10.8 the mean of the quantization error in steady state for each controller is indicated. As expected, controller $C_{3_{DM}}(s)$ has, in general, better noise rejection while $C_{1_{DM}}(s)$ has the worst. Finally according to the above, a suitable choice is controller $C_{3_{DM}}(s)$.

Table 10.5 Settling time response: cool region

Set point (°C)	$C_{1_{DM}}$(s)	$C_{2_{DM}}$(s)	$C_{3_{DM}}$(s)
4 to 2	14.4594	13.0208	14.4161
2 to 8	6.7036	14.2162	11.0374
8 to 6	14.1037	3.6493	8.1321
6 to 2	13.0249	14.4137	12.4692
2 to 4	9.4053	6.7590	15.4354
$J_1(\boldsymbol{\theta})$	8.3280	6.9230	12.500
$J_5(\boldsymbol{\theta})$	0.5361	0.9397	1.0310

Table 10.6 Settling time response: freeze region

Set point (C)	$C_{1_{DM}}$ (s)	$C_{2_{DM}}$ (s)	$C_{3_{DM}}$ (s)
−5 to 0	7.1187	13.0171	14.4854
0 to −2	11.7220	13.4556	9.7237
−2 to −8	8.1127	10.6659	14.4672
−8 to −10	12.2187	14.2245	11.8961
−10 to −6	15.6550	11.2335	14.2329
$J_3(\boldsymbol{\theta})$	9.8420	11.5500	15.5000
$J_6(\boldsymbol{\theta})$	0.7619	1.7710	1.2160

Table 10.7 Mean error quantization in steady state: cool region

Set point (C)	$C_{1_{DM}}$ (s)	$C_{2_{DM}}$ (s)	$C_{3_{DM}}$ (s)
4 to 2	3.3142	0.1361	0.9368
2 to 8	2.3854	2.1204	0.7495
8 to 6	2.0417	1.2882	0.5925
6 to 2	1.6804	1.0251	0.4117
2 to 4	1.3093	1.0395	0.7116

Table 10.8 Mean error quantization in steady state: freeze region

Set point (C)	$C_{1_{DM}}$ (s)	$C_{2_{DM}}$ (s)	$C_{3_{DM}}$ (s)
−5 to 0	1.1428	0.8498	5.2698
0 to −2	6.1262	0.7220	1.7783
−2 to −8	1.9890	1.1336	0.4974
−8 to −10	5.3525	0.9852	0.5232
−10 to −6	1.4772	0.8804	0.4348

10.5 Conclusions

In this chapter, a MOOD procedure using a multiobjective RBDO was performed in order to tune a controller for a Peltier cell device. As commented in Chap. 7, managing reliability in the MOP definition is useful at the MCDM stage, since it is possible to have an idea about performance and risk of failure or probability of degradation. This statement was useful in order to deal with a nonlinear process, using a subset of linear models describing its dynamics.

In the MCDM phase, a subset of three suitable controllers were selected from the approximated Pareto Front in order to go through additional tests with the real system. As expected, differences between real performances and nominal ones appear due to unmodeled components and their effects (noisy measurements for example). Nevertheless, a good selection of design objectives will reduce such effects (including, for example, a design objective related to noise sensibility). Anyway, if the control engineer is looking for a better match between theoretical performances from Pareto Front and real ones from real tests, then it is required to include such effects, as it was followed in Chap. 8.

References

1. Casano G, Piva S (2011) Experimental investigation of the performance of a thermoelectric generator based on peltier cells. Exp Thermal Fluid Sci 35(4):660–669
2. Frangopol DM, Maute K (2003) Life-cycle reliability-based optimization of civil and aerospace structures. Comput Struct 81(7):397–410
3. Konz I, Fernández B, Fernández ML, Pereiro R, Sanz-Medel A (2014) Design and evaluation of a new peltier-cooled laser ablation cell with on-sample temperature control. Analytica Chimica Acta 809:88–96
4. Mannella GA, Carrubba VL, Brucato V (2014) Peltier cells as temperature control elements: experimental characterization and modeling. Appl Thermal Eng 63(1):234–245
5. Morizane K, Ogata T, Morino T, Horiuchi H, Yamaoka G, Hino M, Miura H (2012) A novel thermoelectric cooling device using peltier modules for inducing local hypothermia of the spinal cord: the effect of local electrically controlled cooling for the treatment of spinal cord injuries in conscious rats. Neurosci Res 72(3):279–282
6. Najafi H, Woodbury KA (2013) Optimization of a cooling system based on peltier effect for photovoltaic cells. Solar Energy 91:152–160
7. Reynoso-Meza G, Sánchez HS, Blasco X, Vilanova R (2014) Reliability based multiobjective optimization design procedure for pi controller tuning. In: 19th World congress of the international federation of automatic control, 2014
8. Reynoso-Meza G, Sanchis J, Blasco X, García-Nieto S (2014) Physical programming for preference driven evolutionary multi-objective optimization. Appl Soft Comput 24:341–362

Chapter 11
Multiobjective Optimization Design Procedure for Controller Tuning of a TRMS Process

Abstract In this chapter, the multiobjective optimization design procedure will be used in order to tune the controller of a Twin Rotor MIMO System (TRMS). For such process, a many-objectives optimization instance is tackled using aggregate objective functions. Two different controllers are compared: a decentralized PID structure and a State Space feedback controller.

11.1 Introduction

MOOD procedure, is not just useful for finding a desirable balance of conflictive design objectives for a given controller structure but it might be valuable to understand the trade-off in an overall sense. That is, it could be used to understand better the control problem at hand, and take a more reliable and comfortable decision on the design alternative selected.

In this chapter, such analysis will be performed over a multivariable system, a Twin Rotor MIMO system, comparing two control alternatives (a multivariable PID and a State Space feedback controller). Taking profit of LD tool, it will be concluded which control structure will be used, understanding trade-offs among conflictive objectives, coupling effects and robustness. Evaluating two different control structures will allow us to decide if a complex structure is justifiable for a multivariable process like this.

11.2 Process Description

A nonlinear Twin Rotor MIMO System (TRMS) (See Fig. 11.1a) manufactured by Feedback Instruments,[1] is used. The TRMS is an academic workbench and a useful platform to evaluate control strategies [3–6] due to its complexity and coupling effects. It is a two input, two output system, where two DC motors have control over

[1] http://www.feedback-instruments.com/products/education.

G. Reynoso Meza et al., *Controller Tuning with Evolutionary Multiobjective Optimization*, Intelligent Systems, Control and Automation: Science and Engineering 85, DOI 10.1007/978-3-319-41301-3_11

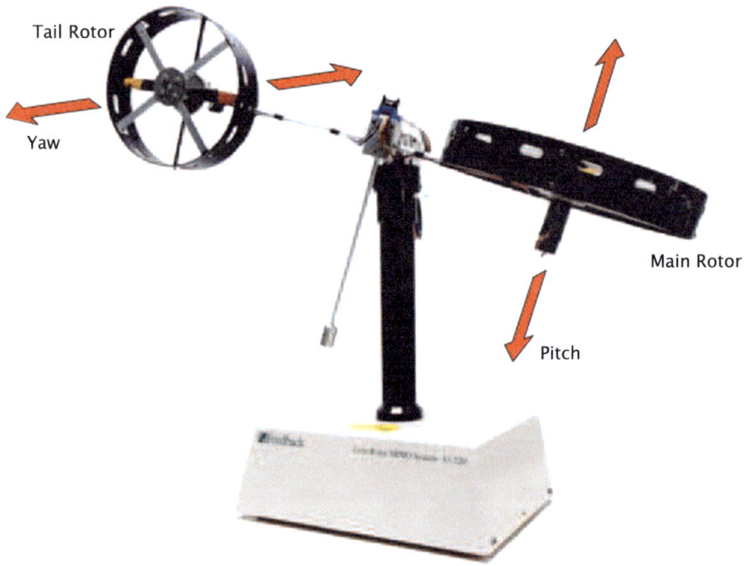

Fig. 11.1 Twin Rotor MIMO System (TRMS) by feedback instruments

two controlled angles. The first one is the vertical (pitch or main) angle controlled by the main rotor and the second one is a horizontal (yaw or tail angle) angle, controlled by the tail rotor (See Fig. 11.1b). Both inputs are normalized in the range $[-1, 1]$ while pitch angle is in the range $[-0.5, 0.5]$ rad and yaw angle in the range $[-3.0, 3.0]$ rad.

The nonlinear model of the system is as follows [1, 2]:

$$\frac{d\alpha_v}{dt} = \Omega_v \tag{11.1}$$

$$\frac{d\Omega_v}{dt} = f_1(\alpha_v, \Omega_v, \alpha_h, \Omega_h, \omega_m, \omega_t, u_m, u_t) \tag{11.2}$$

$$\frac{\omega_m}{dt} = f_2(\bar{\omega}_m, u_m) \tag{11.3}$$

$$\frac{d\alpha_h}{dt} = \Omega_h \tag{11.4}$$

$$\frac{d\Omega_h}{dt} = f_3(\alpha_v, \Omega_v, \alpha_h, \Omega_h, \omega_m, \omega_t, u_m, u_t) \tag{11.5}$$

$$\frac{d\omega_t}{dt} = f_4(\omega_t, u_t) \tag{11.6}$$

where α_v, α_h are the pitch and yaw angles respectively; Ω_v, Ω_h their vertical and horizontal angular velocities and ω_m ω_t the rotational velocities of main and tail rotors. Variables u_m and u_t are the input variables for main and tail rotors respectively. The TRMS is a coupled system, since both rotors produce variations in pitch and yaw displacements. For a detailed explanation of the model, interested readers are invited to consult references [1, 2]. In summary, it is a nonlinear coupled MIMO process.

11.3 The MOOD Approach for Design Concepts Comparison

As commented before, two different control structures will be evaluated in an overall sense with the purpose of having a more general picture on the affordable tradeoff in each instance. Such analysis will give us the answer if it is worth using a complex control structure over a simple one. The control structures will be: a decentralized multivariable PID controller (See Fig. 11.2), and a State Space feedback matrix (See Fig. 11.3). It is expected that a State Space feedback controller gets a better overall performance because more information is used in the closed loop to calculate the control action. Prior to approximating a Pareto Front for such controller, this fact will be justified by using a design concepts comparison with the decentralized PID controller.

To evaluate the performance of a given controller, a Simulink©model with the identified nonlinear model is used. Two simulations of 40 s each one will be carried out:

- Test 1 (T1): Pitch angle setpoint change from 0 to 0.4 rad while yaw setpoint is at 0 rad.
- Test 2 (T2): Yaw angle setpoint change from 0 to 2.4 rad while pitch setpoint is at 0 rad.

The same design objectives as defined in [5] will be used. The objectives take into consideration time performance $J_1(\boldsymbol{\theta})$ and coupling effects $J_3(\boldsymbol{\theta})$ using ratios of IAE

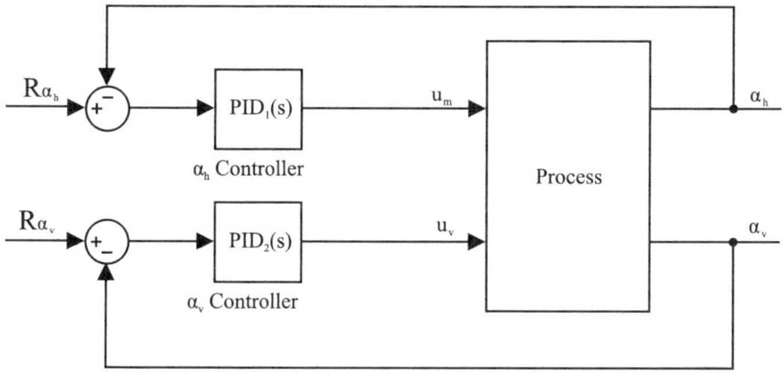

Fig. 11.2 PID control loops. $PID_1(s) = Kc_1 \frac{Td_1 s^2 + s + 1/Ti_1}{s}$ and $PID_2(s) = Kc_2 \frac{Td_2 s^2 + s + 1/Ti_2}{s}$

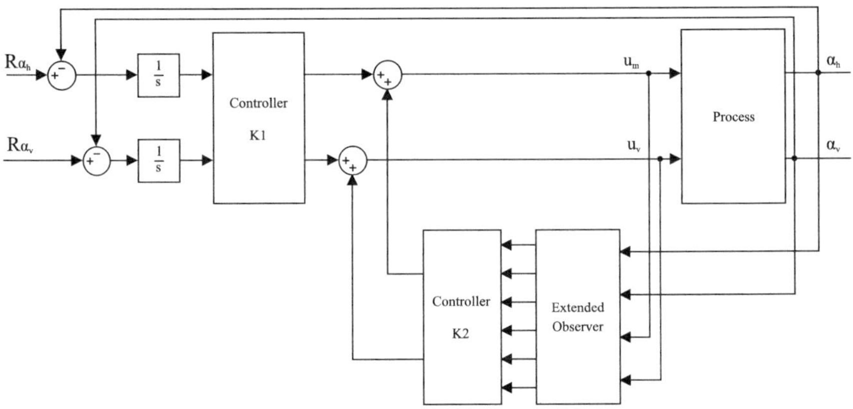

Fig. 11.3 State space control loop with extended observer. $K1$ is a 2×2 matrix and $K2$ a 2×6 matrix

(Eq. 2.9) (for pitch and yaw); and the usage of control action $J_2(\boldsymbol{\theta})$ by means of TV ratios (Eq. 2.18) (for main and tail rotor). Such ratios will be calculated with tests T1 and T2:

- $J_1(\boldsymbol{\theta})$: aggregate objective function of the normalized IAE (Eq. 2.9) for pitch and yaw angles, in order to get a desired setpoint.

$$J_1(\boldsymbol{\theta}) = T_s \begin{bmatrix} 1 \cdot \max\left(\frac{IAE^{pitch,T1}(\boldsymbol{\theta})}{0.4}, \frac{IAE^{yaw,T2}(\boldsymbol{\theta})}{2.4}\right) + \\ 0.1 \cdot \min\left(\frac{IAE^{pitch,T1}(\boldsymbol{\theta})}{0.4}, \frac{IAE^{yaw,T2}(\boldsymbol{\theta})}{2.4}\right) \end{bmatrix} \tag{11.7}$$

- $J_2(\boldsymbol{\theta})$: aggregate objective function of the normalized total variation (TV) of control action.

$$J_2(\boldsymbol{\theta}) = \begin{bmatrix} 1 \cdot \max\left(TV^{Main,T1}(\boldsymbol{\theta}) + TV^{Main,T2}(\boldsymbol{\theta}), TV^{Tail,T1}(\boldsymbol{\theta}) + TV^{Tail,T2}(\boldsymbol{\theta})\right) + \\ 0.1 \cdot \min\left(TV^{Main,T1}(\boldsymbol{\theta}) + TV^{Main,T2}(\boldsymbol{\theta}), TV^{Tail,T1}(\boldsymbol{\theta}) + TV^{Tail,T2}(\boldsymbol{\theta})\right) \end{bmatrix} \tag{11.8}$$

- $J_3(\boldsymbol{\theta})$: aggregate objective function of the normalized IAE (Eq. 2.9) for pitch and yaw angles due to coupling effects.

$$J_3(\boldsymbol{\theta}) = T_s \begin{bmatrix} 1 \cdot \max\left(\frac{IAE^{yaw,T1}(\boldsymbol{\theta})}{1}, \frac{IAE^{pitch,T2}(\boldsymbol{\theta})}{6}\right) + \\ 0.1 \cdot \min\left(\frac{IAE^{yaw,T1}(\boldsymbol{\theta})}{1}, \frac{IAE^{pitch,T2}(\boldsymbol{\theta})}{6}\right) \end{bmatrix}. \tag{11.9}$$

Such definitions bring the convenience of agglutinating similar design objectives and it will be convenient in order to analyse the two design concepts under consideration. Pertinency bounds are defined according to results in [5]. For the case of PID controller, decision variables are $\boldsymbol{\theta} = [Kc_1, Ti_1, Td_1, Kc_2, Ti_2, Td_2]$. Therefore, the MOP statement at hand is:

$$\min_{\boldsymbol{\theta}} \boldsymbol{J}(\boldsymbol{\theta}) = [J_1(\boldsymbol{\theta}), J_2(\boldsymbol{\theta}), J_3(\boldsymbol{\theta})] \tag{11.10}$$

subject to[2]

$$0 \leq Kc_1, Kc_2 \leq 20 \tag{11.11}$$
$$0 \leq Ti_1, Ti_2 \leq 200 \tag{11.12}$$
$$0 \leq Td_1, Td_2 \leq 20 \tag{11.13}$$
$$J_1(\boldsymbol{\theta}) \leq 8 \tag{11.14}$$
$$J_2(\boldsymbol{\theta}) \leq 2 \tag{11.15}$$
$$J_3(\boldsymbol{\theta}) \leq 1 \tag{11.16}$$
$$\text{Stable in closed loop.} \tag{11.17}$$

For the state space controller, decision variables are the elements of the feedback gain matrix K, $\boldsymbol{\theta} = [K1^{11}, \cdots, K1^{22}, K2^{11}, \cdots, K2^{26}]$. Therefore, the MOP statement at hand is:

$$\min_{\boldsymbol{\theta}} \boldsymbol{J}(\boldsymbol{\theta}) = [J_1(\boldsymbol{\theta}), J_2(\boldsymbol{\theta}), J_3(\boldsymbol{\theta})] \tag{11.18}$$

subject to:

$$-20 \leq K1^{11}, \cdots, K1^{22}, K2^{11}, \cdots, K2^{26} \leq 20 \tag{11.19}$$
$$J_1(\boldsymbol{\theta}) \leq 8 \tag{11.20}$$
$$J_2(\boldsymbol{\theta}) \leq 2 \tag{11.21}$$
$$J_3(\boldsymbol{\theta}) \leq 1 \tag{11.22}$$
$$\text{Stable in closed loop.} \tag{11.23}$$

[2]Integral action of the PID is disabled when Ti_1 or Ti_2 equal 0.

Table 11.1 Parameters used for sp-MODE. Further details in Chap. 3

		Parameter	Value
sp-MODE		**Evolutionary mechanism**	
	DE	F (Scaling factor)	0.5
		Cr (Crossover rate)	0.9
		N_p (Population)	20
		Pruning mechanism	
		$\boldsymbol{\beta}^{\epsilon}$ (Arcs)	15

In both cases, nominal stability will be evaluated with a linearized model of the system. Given that a simple pertinency mechanism is required, the sp-MODE algorithm will be used, with the parameters depicted in Table 11.1.

In Fig. 11.4 a design concept comparison between PID control (red circles) and State Space feedback matrix (blue diamonds) is given. It can be observed that objective J_1 is in conflict with objectives J_2 and J_3. In Fig. 11.4b we can appreciate that PID control dominates in certain areas, while it is being dominated in others.

It is interesting to appreciate how tendencies change in design objective $J_1(\boldsymbol{\theta})$: for values below ≈ 4, PID control seems to be the better option, while for values above ≈ 4 the State Space controller performs better. This also could mean that, for the function evaluation budget, it is easier to tune the PID control.

A designer might be tempted to pick a PID control with this information, nevertheless the information from design objectives 2 and 3 is also relevant. From design objective 3 (coupling effects) it is possible to appreciate that State Space controller performs better, meaning that it is more capable of dealing with coupling effects (which in fact, is expected since the controller uses the state information of the system). Regarding $J_1(\boldsymbol{\theta})$ (control action), a State Space matrix has a better usage of control action. It can be observed that the range $J_2 \in [0.5, 0.8]$ is not covered by PID control, only a state feedback controller can obtain these values.

Therefore, given the above comments it is justified to use a State Space controller instead of a PID controller. That is, it is justified to use a complex controller for this process.

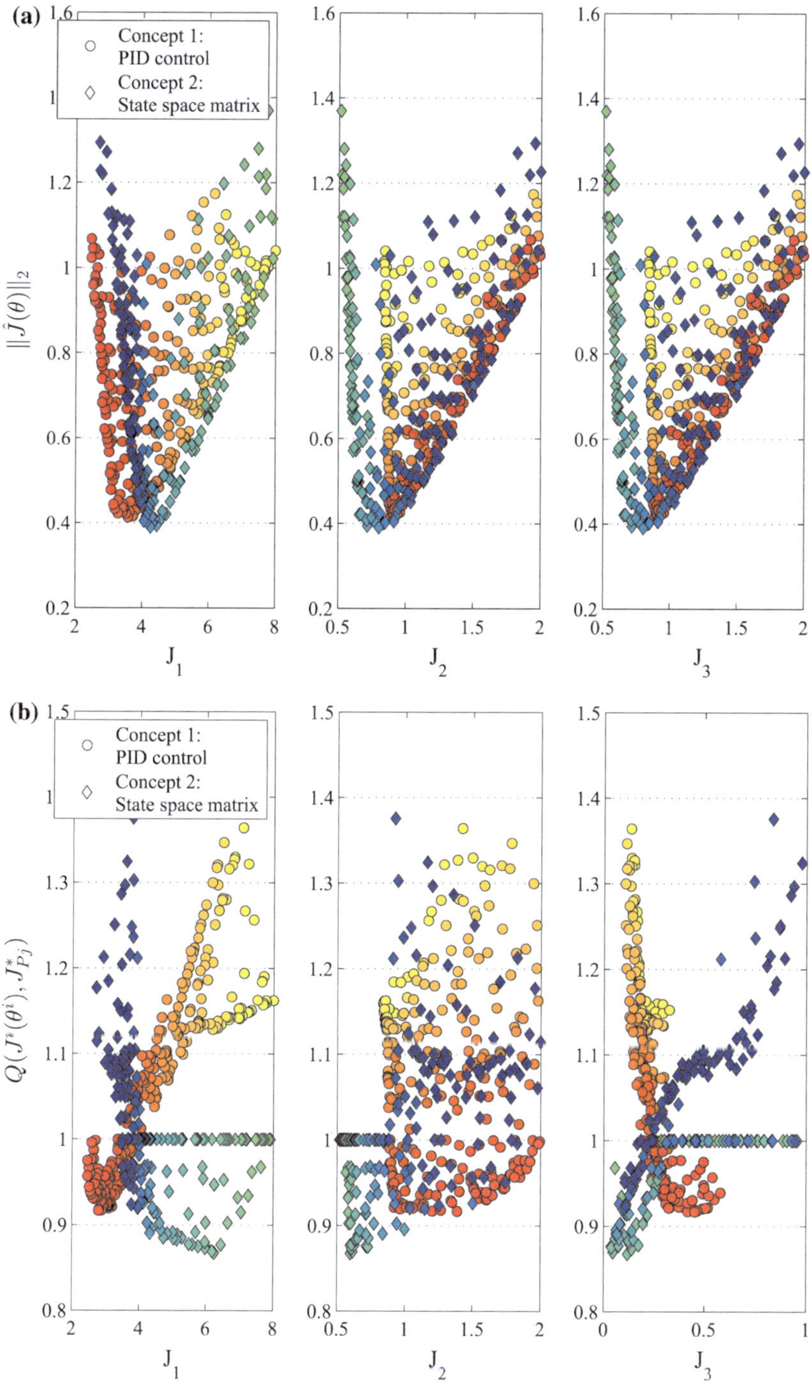

Fig. 11.4 Design concepts comparison using LD and quatilty indicator Q (see Chap. 3). **a** LD. **b** Q

11.4 The MOOD Approach for Controller Tuning

The latter section concluded that a space state as control structure is justifiable. Nevertheless additional design objectives are required in order to guarantee useful solutions when the controller is implemented in the real process. For this purpose, two new design objectives are incorporated: one for noise performance $J_4(\boldsymbol{\theta})$ and one for robust performance $J_5(\boldsymbol{\theta})$.

$$J_4(\boldsymbol{\theta}) = \boldsymbol{\theta} * \boldsymbol{\theta}^T, \tag{11.24}$$

$$J_5(\boldsymbol{\theta}) = \sup_{\omega} \left(\mathcal{T}(j\omega)W(j\omega)\right), \omega \in [10^{-2} \ldots 10^2], \tag{11.25}$$

where $\mathcal{T}(j\omega)$ is the complementary sensitivity function and function $W(j\omega) = \frac{0.7j\omega+2}{j\omega+1.1}$ will be used. Therefore, the MOP at hand is:

$$\min_{\boldsymbol{\theta}} \boldsymbol{J}(\boldsymbol{\theta}) = [J_1(\boldsymbol{\theta}), \cdots J_5(\boldsymbol{\theta})] \tag{11.26}$$

subject to:

$$-20 \leq K1^{11}, \cdots, K1^{22}, K2^{11}, \cdots, K2^{26} \leq 20 \tag{11.27}$$

$$J_1(\boldsymbol{\theta}) \leq 8 \tag{11.28}$$

$$J_2(\boldsymbol{\theta}) \leq 2 \tag{11.29}$$

$$J_3(\boldsymbol{\theta}) \leq 1 \tag{11.30}$$

$$J_4(\boldsymbol{\theta}) \leq 10 \tag{11.31}$$

$$J_5(\boldsymbol{\theta}) \leq 5 \tag{11.32}$$

$$\text{Stable in closed loop.} \tag{11.33}$$

The sp-MODE algorithm will be used, with same parameters of Table 11.1. In Fig. 11.5 the approximated Pareto Set and Pareto Front are represented with LD. In order to proceed with the MCDM stage, some additional preferences have been considered: a robust solution is preferred due to implementation issues, then solutions with lower J_5 have priority. For the remaining objectives, decoupling behavior is important (then low J_3 is selected), low time response (low J_1), average noise rejection (J_4 in the middle of the scale) and control action economy have the less priority (high J_2 is allowed). To verify the final solution comparing with the rest of the Pareto solutions, additional information from time responses is given in Figs. 11.6 and 11.7. This additional information gives an insight about the (subjective) quality of the time performance for each one of the controllers. Finally, selected controller K_{DM} is indicated for implementation.

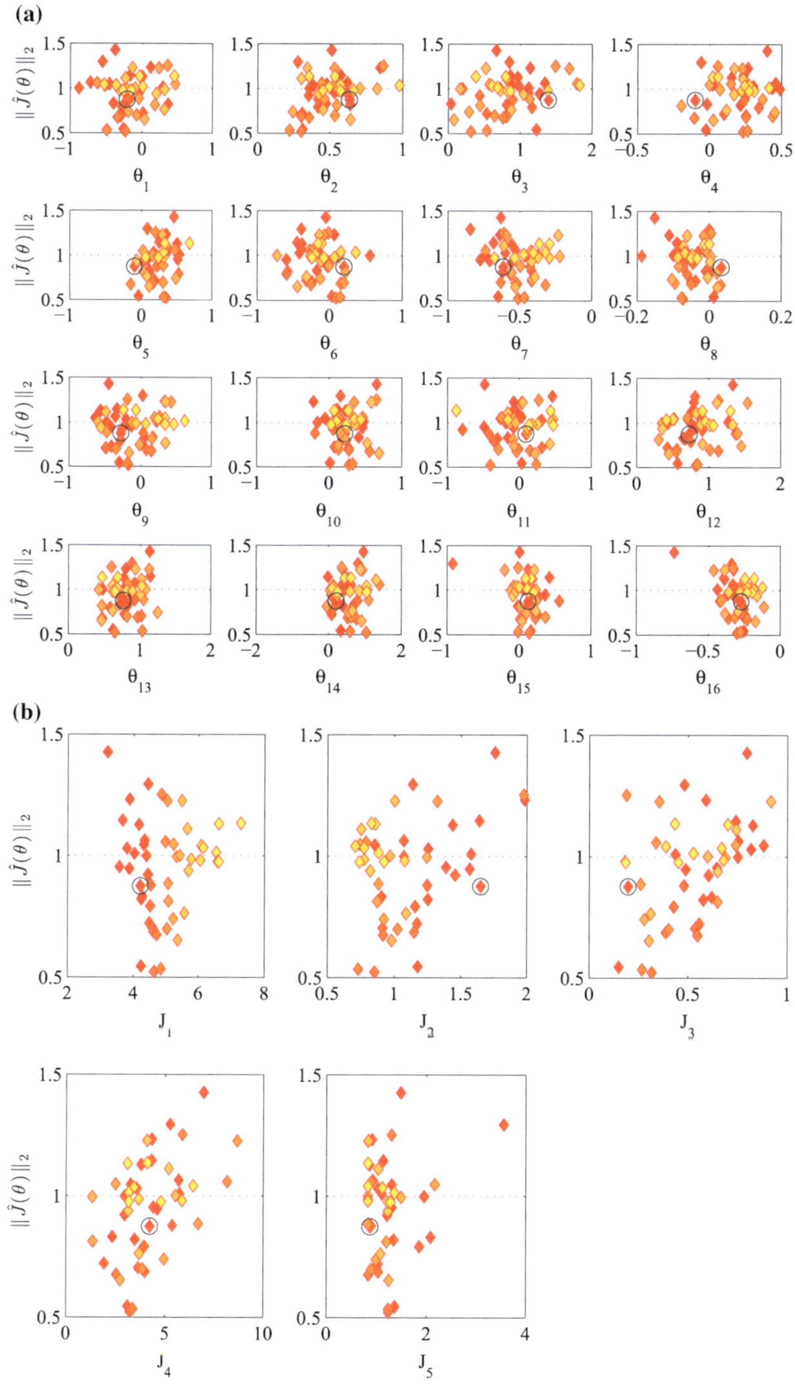

Fig. 11.5 Pareto set and front approximations. By means of ○, controller selected K_{DM}, is indicated. **a** Pareto Set. **b** Pareto Front

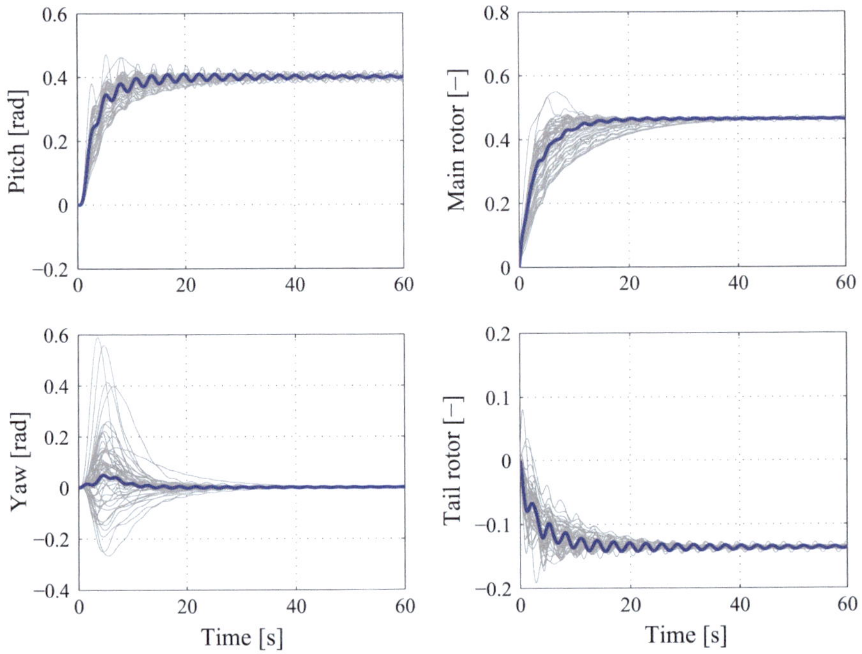

Fig. 11.6 Performance on test T1 of the approximated pareto set. Closed loop response obtained with K_{DM} controller (*bold*)

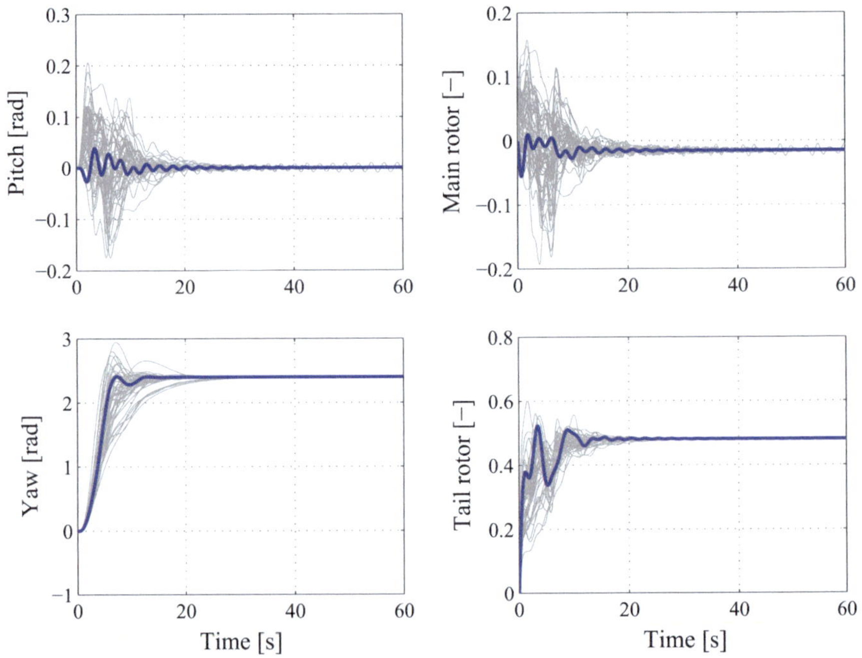

Fig. 11.7 Performance on test T2 of the approximated pareto set. Close loop response obtained with K_{DM} controller (*bold*)

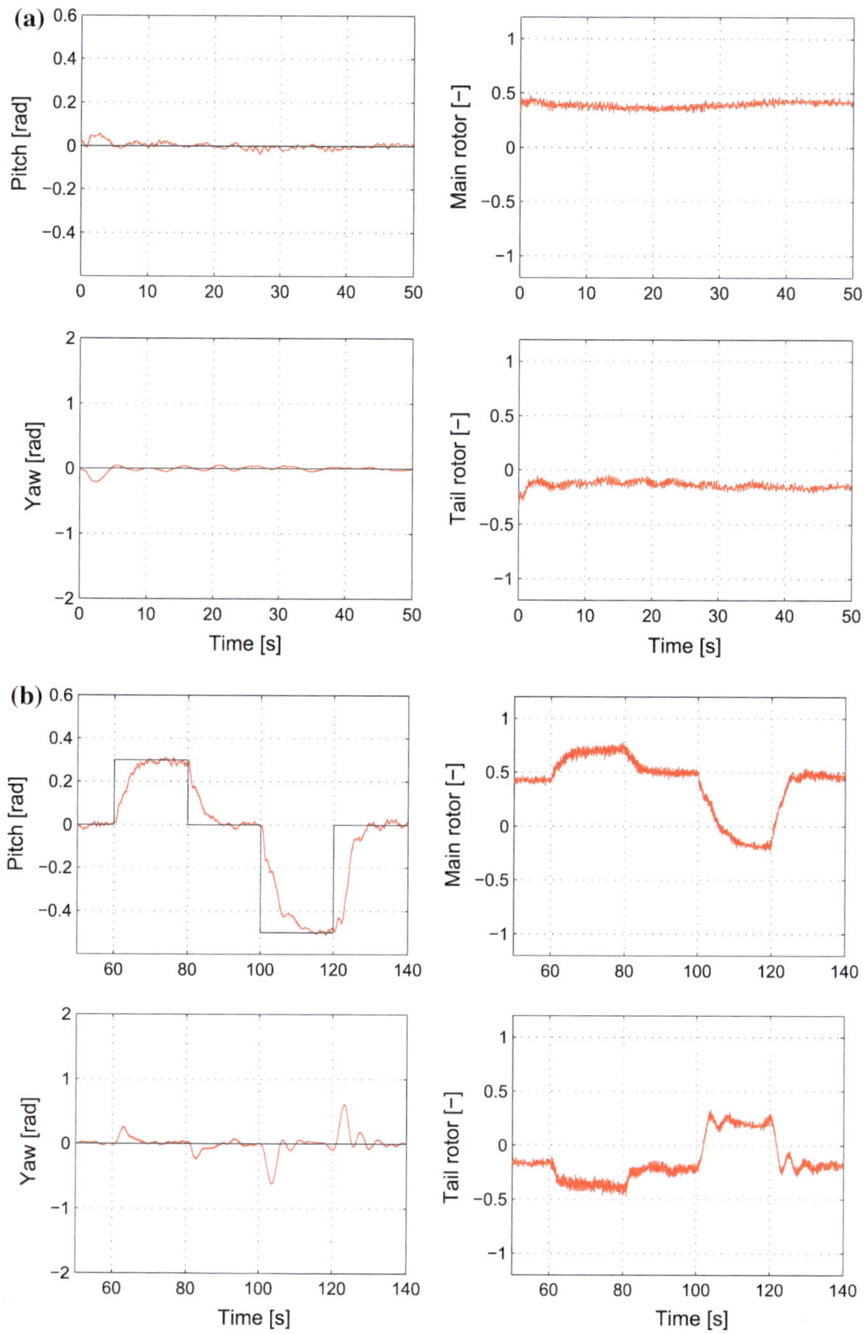

Fig. 11.8 Test A: set point for *Pitch* = 0 rad and *Yaw* = 0 rad respectively. Test B: a sequence of steps in set point for *Pitch* whilst setpoint for *Yaw* = 0 rad. **a** A. **b** B

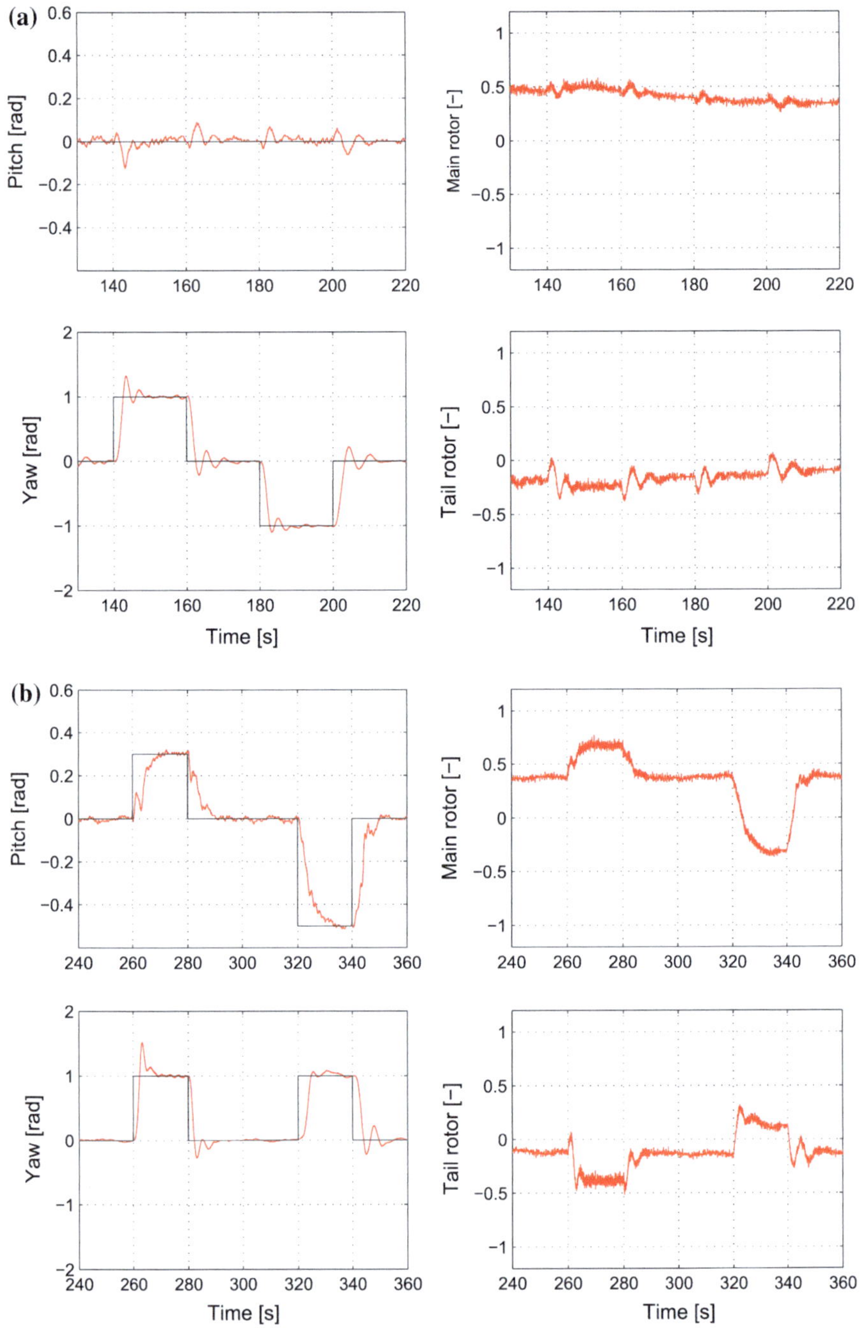

Fig. 11.9 Test C: A sequence of steps in setpoint for *Yaw* whilst setpoint for for *Pitch* = 0 rad. Test D: a sequence of simultaneous steps in setpoint for *Pitch* and *Yaw* respectively. **a** C. **b** D

11.5 Control Tests

The selected controller K_{DM} is implemented in the TRMS control system. Performances of the controller are shown in Figs. 11.8 and 11.9 for different setpoint changes. Notice that such a controller fulfills expectations about the control loop performance.

11.6 Conclusions

In this chapter, Pareto Fronts for PID and State Space controllers were approximated for a TRMS. As in Chap. 7, an overall comparison (instead of punctual) of the achievable tradeoff between two different control structures was performed. With such comparison, it was possible to identify strengths of one controller structure (the more complex) over the other (the simplest). In this way the control engineer can evaluate if such improvement on performance compensates using one controller over the other. After such design concepts comparison, the regular MCDM process is carried out using additional information from closed loop time responses, in order to ponder tradeoff of each controller.

References

1. Carrillo-Ahumada J, Reynoso-Meza G, García-Nieto S, Sanchis J, García-Alvarado M (2015) Sintonización de controladores pareto-óptimo robustos para sistemas multivariables. aplicación en un helicóptero de 2 grados de libertad. Revista Iberoamericana de Automática e Informática Industrial RIAI 12(2):177–188
2. Gabriel C (2009) Modelling, simulation and control of a twin rotor mimo-system. Master's thesis, Universitat Politcnica de Valncia. http://personales.upv.es/gilreyme/mood4ct/files/TRMS.zip
3. Juang J-G, Lin R-W, Liu W-K (2008) Comparison of classical control and intelligent control for a mimo system. Appl Math Comput 205(2):778–791. Special Issue on Advanced Intelligent Computing Theory and Methodology in Applied Mathematics and Computation
4. Montes de Oca S, Puig V, Witczak M, Quevedo J Fault-tolerant control of a two-degree of freedom helicopter using LPV techniques 1204–1209
5. Reynoso-Meza G, García-Nieto S, Sanchis J, Blasco X (2013) Controller tuning using multiobjective optimization algorithms: a global tuning framework. IEEE Trans Control Syst Technol 21(2):445–458
6. Wen P, Lu T-W (2008) Decoupling control of a twin rotor mimo system using robust deadbeat control technique. IET Control Theory Appl 2(11):999–1007

Chapter 12
Multiobjective Optimization Design Procedure for an Aircraft's Flight Control System

Abstract In this chapter, the multiobjective optimization design procedure will be used to tune the autopilot controllers for an autonomous Kadett© aircraft. For this aim, a multivariable PI controller is defined, and a many-objectives optimization instance is tackled using designer preferences. After the multicriteria decision making stage, the selected controller is implemented and evaluated in a real flight test.

12.1 Introduction

Nowadays, Unmanned Aerial Vehicles (UAV's) are an emerging and strategic research topic [13] with great potential in commercial and civil applications as monitoring (pipes, crop fields, forest, weather), sensing and recording (pollution, vigilance) [3]. One of the most important devices is the Flight Control System (FCS), which provides the desired level of autonomy to the vehicle. Several alternatives for control algorithms have been used in the FCS of UAVs, in order to provide the autonomy level required to accomplish their tasks. For example, proportional-integral-derivative (PID) controllers [6], linear quadratic regulators (LQR) [14], fuzzy logic techniques [4], artificial neural networks [8], adaptive control [12] and predictive control [2] have been extensively used for such purpose. Nevertheless, new control techniques and procedures are still required in order to improve the mission performance of UAVs [1].

In this chapter, three different control loops for an UAV will be tuned. Two of them are cascade loops for heading and altitude control, while the third is a simple control loop for velocity. In order to accomplish such task, many-objective optimization problem will be defined and a reference controller is used to determine the pertinent region of the objective space.

G. Reynoso Meza et al., *Controller Tuning with Evolutionary Multiobjective Optimization*, Intelligent Systems, Control and Automation: Science and Engineering 85, DOI 10.1007/978-3-319-41301-3_12

12.2 Process Description

In Fig. 12.1 the aircraft for test and validation is presented. As the main component of the flight platform, a Kadett© 2400 aircraft, manufactured by Graupner,[1] is used. It is a light weight airframe with some features that make it suitable for the purposes of this research. Some of those characteristics are:

- Wing span of 2.4 [m].
- Wing area of 0.9 [m^2].
- Weight/area ratio of 49 [$\frac{g}{dm^2}$].
- Free volume of 16.5 [l].

Fig. 12.1 Kadett© aircraft (by Graupner)

[1] http://www.graupner.de/en/.

During flight, three control surfaces are provided: tail[2] rudder u_{RU}, elevators u_E and ailerons u_A. For propulsion u_T, a brushless engine of alternating current is integrated fed by two LiPo[3] batteries through a frequency converter. Alike servomotors, converters are controlled by sending Pulse Width Modulated (PWM) signals as commands (control actions are sent from the FCS). The loop is closed by a GPS-AHRS IG500N unit,[4] which includes accelerometers, gyroscopes and magnetometers. Its Kalman filter is capable of mixing the information coming from those sensors in order to offer precise measurements of position, orientation, linear and angular speeds and accelerations, in the three aircraft body-axes. In [9] is presented this platform with more details together with the results of some flight tests.

A general non linear model [10, 11] for an aircraft like this is given by:

$$\overrightarrow{F_A} + \overrightarrow{F_T} + \overrightarrow{F_G} = m\dot{\overrightarrow{V}} + \overrightarrow{\omega} \times \overrightarrow{V} \tag{12.1}$$

$$\overrightarrow{M_A} = I\dot{\overrightarrow{\omega}} + \overrightarrow{\omega} \times I\overrightarrow{\omega} \tag{12.2}$$

where F_A is the aerodynamic force; $F_T(u_T)$ the force applied by the motor; F_G is the gravitational force; $\overrightarrow{V}$ and $\overrightarrow{\omega}$ are the linear and angular velocities respectively; $\overrightarrow{M_A}$ the aerodynamic torque and m and I the mass and the inertia tensor of the aircraft respectively. Special attention deserve $\overrightarrow{F_A}$ and $\overrightarrow{M_A}$, which depend on the so called aerodynamic coefficients $C_X(u_{A,E})$, $C_Y(u_{A,E})$, $C_Z(u_{A,E})$, $C_l(u_{A,E})$, $C_m(u_{A,E})$ and $C_n(u_{A,E})$:

$$\overrightarrow{F_A} = \overline{q}S \begin{bmatrix} C_X \\ C_Y \\ C_Z \end{bmatrix} \tag{12.3}$$

$$\overrightarrow{M_A} = \overline{q}S \begin{bmatrix} bC_l \\ \overline{c}C_m \\ bC_n \end{bmatrix} \tag{12.4}$$

where S, b and $\overline{c}$ are constructive constants of the aircraft and $\overline{q}$ is the dynamic pressure of the air. Such coefficients are functions that correlate forces and torques to system variables. Our model is taken from [11], where aerodynamic coefficients take the polynomial form stated in [5] and were calculated using MOOD techniques.

Basically, the FCS should manipulate yaw, pitch and roll angles (see Fig. 12.2) in order to guarantee sustainability for the desired flight task. For such purpose, two cascade loops are defined.

---

[2] Tail rudder control is obtained as a ratio control from ailerons control: $u_{RU} = 0.25u_A$.

[3] Lithium polymer battery.

[4] http://www.sbg-systems.com/products/ig500n-miniature-ins-gps.

Fig. 12.2 Yaw, pitch and roll angles

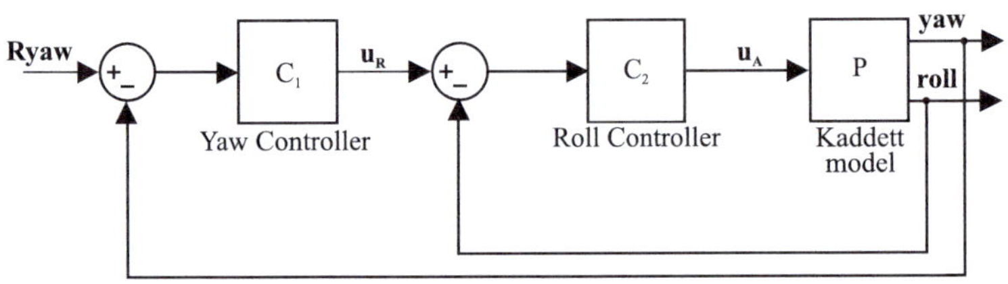

Fig. 12.3 Cascade loop for yaw control

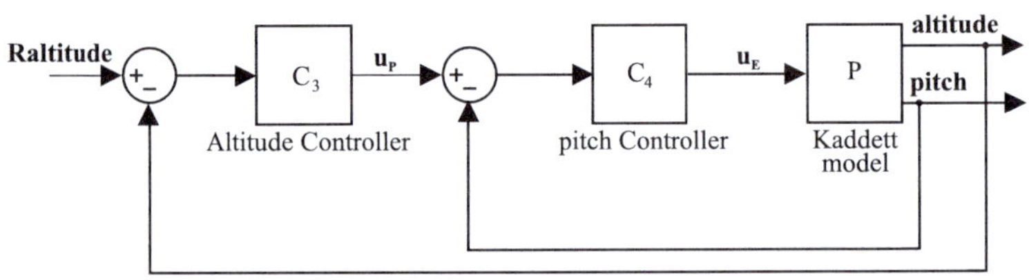

Fig. 12.4 Cascade loop for altitude control

Fig. 12.5 Velocity control loop

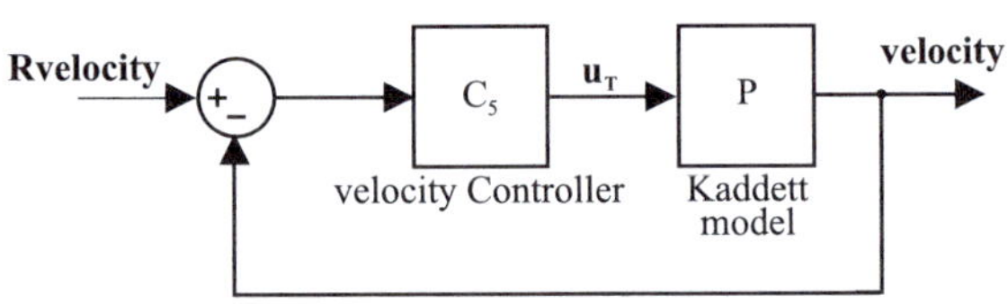

The first cascade loop (Fig. 12.3) keeps under the desired reference the yaw angle (or heading) by manipulating roll reference and ailerons deflections. The second cascade loop (Fig. 12.4) keeps under the desired reference the altitude, by manipulating pitch reference and elevators deflections.

An additional control loop Fig. 12.5 is used for control velocity, by manipulating the motor throttle.

Thus, a total of five controllers need to be tuned, in order to guarantee the expected performance of the aircraft. It will use a total of five proportional-integral (PI) controllers:

$$C_j(s) = Kc_j\left(1 + \frac{1}{Ti_j \cdot s}\right) \quad j \in [1 \ldots 5]. \tag{12.5}$$

12.3 The MOOD Approach

Simulink© model of the Kadett 2400 will serves us to test controller's performance when simultaneously setpoint changes in altitude and yaw are applied. With this, autopilot's ability to reach a desired aircraft configuration, as well as keeping the aircraft's sustainability via throttle control are evaluated. Design objectives stated are:

- $J_1(\boldsymbol{\theta})$: Settling time for heading (yaw) at $\pm 2\,\%$.

$$J_1(\boldsymbol{\theta}) = J_{t_{98\,\%}}(\boldsymbol{\theta}) \tag{12.6}$$

- $J_2(\boldsymbol{\theta})$: Settling time for altitude at $\pm 2\,\%$.

$$J_2(\boldsymbol{\theta}) = J_{t_{98\,\%}}(\boldsymbol{\theta}) \tag{12.7}$$

- $J_3(\boldsymbol{\theta})$: Throttle's total variation of control action (Eq. 2.18).

$$J_3(\boldsymbol{\theta}) = \int_{t=t_0}^{t_f} \left|\frac{du_T}{dt}\right| dt \tag{12.8}$$

- $J_4(\boldsymbol{\theta})$: Aileron's total variation of control action (Eq. 2.18).

$$J_4(\boldsymbol{\theta}) = \int_{t=t_0}^{t_f} \left|\frac{du_A}{dt}\right| dt \tag{12.9}$$

- $J_5(\boldsymbol{\theta})$: Elevator's total variation of control action (Eq. 2.18).

$$J_5(\boldsymbol{\theta}) = \int_{t=t_0}^{t_f} \left|\frac{du_E}{dt}\right| dt \tag{12.10}$$

- $J_6(\boldsymbol{\theta})$: Roll's total variation of control action (Eq. 2.18).

$$J_6(\boldsymbol{\theta}) = \int_{t=t_0}^{t_f} \left| \frac{du_R}{dt} \right| dt \tag{12.11}$$

- $J_7(\boldsymbol{\theta})$: Pitch's total variation of control action (Eq. 2.18).

$$J_7(\boldsymbol{\theta}) = \int_{t=t_0}^{T_f} \left| \frac{du_P}{dt} \right| dt. \tag{12.12}$$

Design objectives $J_1(\boldsymbol{\theta})$ and $J_2(\boldsymbol{\theta})$ are stated for performance while $J_3(\boldsymbol{\theta})$ to $J_7(\boldsymbol{\theta})$ for robust performance. Controllers with the lowest effort (and fulfilling the flight task) are used in order to avoid aggressive control actions which might compromise the aircraft integrity (performing threatening maneuvers) and its payload (oscillating control actions). A preference matrix is shown in Table 12.1, using an available controller for reference $\boldsymbol{\theta}_{ref}$. Notice that $J_3(\boldsymbol{\theta})$ to $J_7(\boldsymbol{\theta})$ preferences use the reference case controller $\boldsymbol{\theta}_{ref}$ to provide some meaning to values obtained using Eq. 2.18 since these objectives, by themselves, do not provide the same level of interpretability as time domain objectives ($J_1(\boldsymbol{\theta})$ and $J_2(\boldsymbol{\theta})$) for which is easy to state preferences. This idea has been exposed in [7].

Therefore, the MOP under consideration is:

$$\min_{\boldsymbol{\theta}} \boldsymbol{J}(\boldsymbol{\theta}) = [J_1(\boldsymbol{\theta}), \ldots, J_7(\boldsymbol{\theta})] \tag{12.13}$$

$$\boldsymbol{\theta} = [Kc_1, Ti_1, \cdots, Kc_5, Ti_5] \tag{12.14}$$

subject to:

$$0 \leq \quad Kc_{1,\cdots,5} \quad \leq 5 \tag{12.15}$$

$$0 < \quad Ti_{1,\cdots,5} \quad \leq 50 \tag{12.16}$$

$$\text{Subject to preferences} \tag{12.17}$$

According to this, the MOO process is performed with sp-MODE-II. Design objectives for optimization are $J_1(\boldsymbol{\theta})$ to $J_7(\boldsymbol{\theta})$ but only $J_1(\boldsymbol{\theta})$ and $J_2(\boldsymbol{\theta})$ are used in the pruning mechanism. This means that, while all the design objectives are considered in the MOO process and used to calculate the GPP index in the pruning mechanism of the sp-MODE-II algorithm, only the first two (the most interpretable) are used to partition the objective space. Parameters used for optimization are depicted in Table 12.2.

Table 12.1 Preference matrix. Five preference ranges have been defined relative to $\boldsymbol{\theta}_{ref}$: *Highly Desirable* (HD), *Desirable* (D), *Tolerable* (T) *Undesirable* (U) and *Highly Undesirable* (HU)

Preference matrix $\mathfrak{P}$						
Objective	← HD →	← D →	← T →	← U →	← HU →	
	J_q^0	J_q^1	J_q^2	J_q^3	J_q^4	J_q^5
$J_1(\boldsymbol{\theta})$ (s)	10	15	20	25	50	100
$J_2(\boldsymbol{\theta})$ (s)	10	20	30	40	80	160
$J_3(\boldsymbol{\theta})$ (–)	$0.7 . J_3(\boldsymbol{\theta_{ref}})$	$0.8 . J_3(\boldsymbol{\theta_{ref}})$	$0.9 . J_3(\boldsymbol{\theta_{ref}})$	$1.1 . J_3(\boldsymbol{\theta_{ref}})$	$1.2 . J_3(\boldsymbol{\theta_{ref}})$	$1.4 . J_3(\boldsymbol{\theta_{ref}})$
⋮	⋮	⋮	⋮	⋮	⋮	⋮
$J_7(\boldsymbol{\theta})$ (–)	$0.7 . J_7(\boldsymbol{\theta_{ref}})$	$0.8 . J_7(\boldsymbol{\theta_{ref}})$	$0.9 . J_7(\boldsymbol{\theta_{ref}})$	$1.1 . J_7(\boldsymbol{\theta_{ref}})$	$1.2 . J_7(\boldsymbol{\theta_{ref}})$	$1.4 . J_7(\boldsymbol{\theta_{ref}})$

Table 12.2 Parameters used for sp-MODE-II. Further details in Chap. 3

	Parameter	**Value**
	Evolutionary mechanism	
sp-MODE-II / sp-MODE / DE	F (Scaling factor)	0.5
sp-MODE-II / sp-MODE / DE	Cr (Crossover rate)	0.9
sp-MODE-II / sp-MODE / DE	N_p (Population)	50
	Pruning mechanism	
sp-MODE-II / sp-MODE	$\boldsymbol{\beta}_\epsilon$ (Arcs)	20
	Pertinency mechanism	
sp-MODE-II	J_{GPP}^{max}	$J_{GPP}([25, 40, 1.1 \cdot J_3(\theta_{ref}), \cdots, 1.1 \cdot J_7(\theta_{ref})])$
sp-MODE-II	$\overline{car(J_P^*)}$	20

In Fig. 12.6 the approximated Pareto Set and Front are depicted whilst their time responses are shown in Fig. 12.7. Notice that the approximated set of controllers perform better than the reference controller $\boldsymbol{\theta}_{ref}$. After analyzing such information, controllers $\boldsymbol{\theta}_{DM}$ are selected (indicated with a star in the figure) due to its smoothness in control action, mainly in the lower control loops (aileron and elevator).

12.4 Controllers Performance in a Real Flight Mission

After validation in a *Hardware in the loop* (HIL) platform, selected controller is ready to be implemented and evaluated in a real flight mission. Such mission comprises supervising four *waypoints*. Each waypoint consists in a vector of latitude, longitude and altitude (See Table 12.3) which are managed by a reference manager embedded into the FCS. The reference manager computes the setpoint values for yaw, altitude and velocity control loops. Performance of selected controller $\boldsymbol{\theta}_{DM}$ accomplish the flight mission defined in Table 12.3 is depicted in Fig. 12.8. Inner loops performance are shown in Fig. 12.9 where as it can be noticed, a successful control structure was tuned in order to fulfil this flight task.

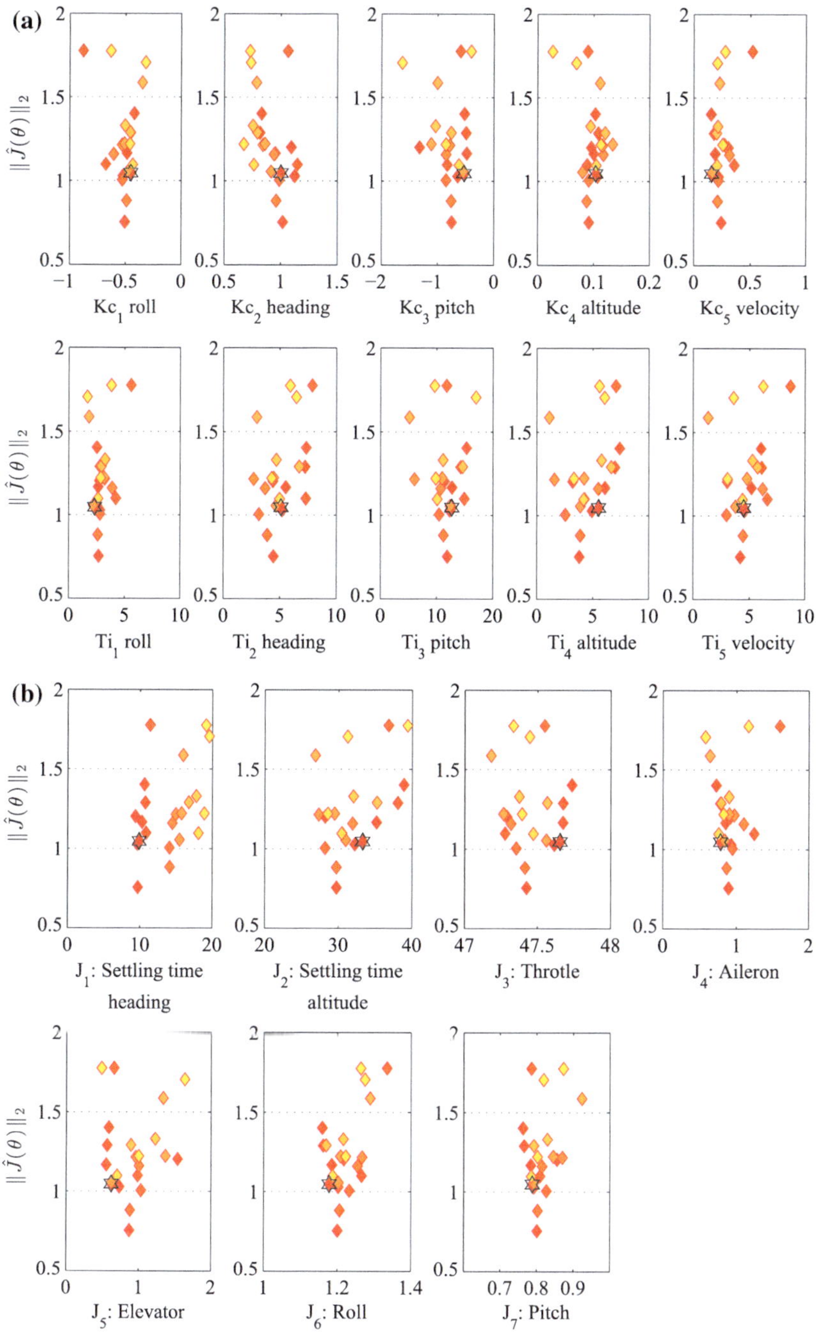

Fig. 12.6 Pareto set and front approximations. By means of ⋆, selected $\boldsymbol{\theta}_{DM}$ controller is represented. **a** Pareto set. **b** Pareto front

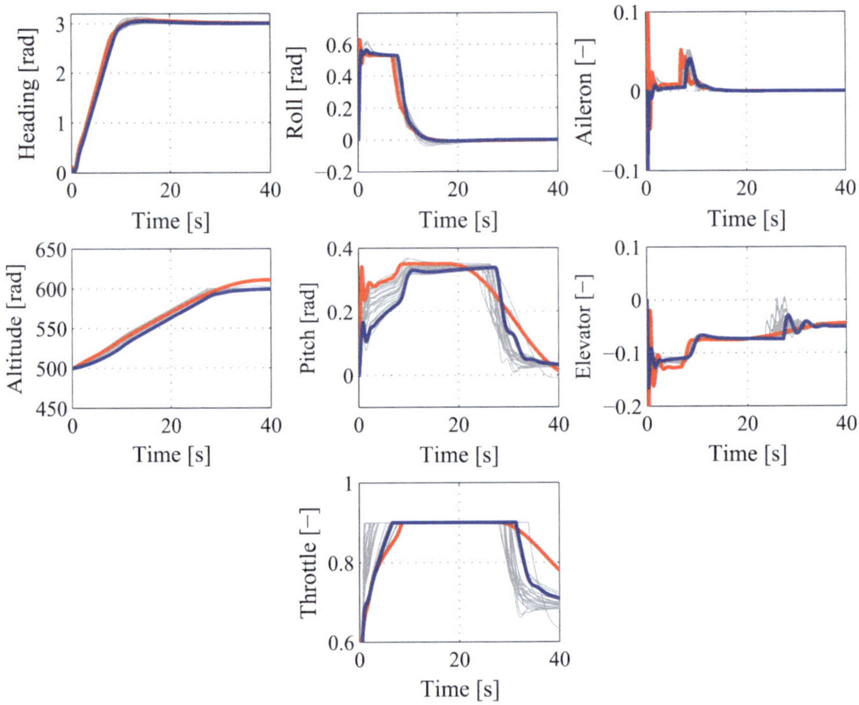

Fig. 12.7 Time performance of the approximated pareto set. The response of the $\boldsymbol{\theta}_{ref}$ and $\boldsymbol{\theta}_{DM}$ are represented in *red* and *blue colors* respectively

Table 12.3 Waypoints definition

	Latitude (°)	Longitude (°)	Altitude (m)
Waypoint 1	39.496126	−0.624117	300
Waypoint 2	39.498278	−0.620856	350
Waypoint 3	39.499239	−0.623087	350
Waypoint 4	39.497517	−0.626435	300

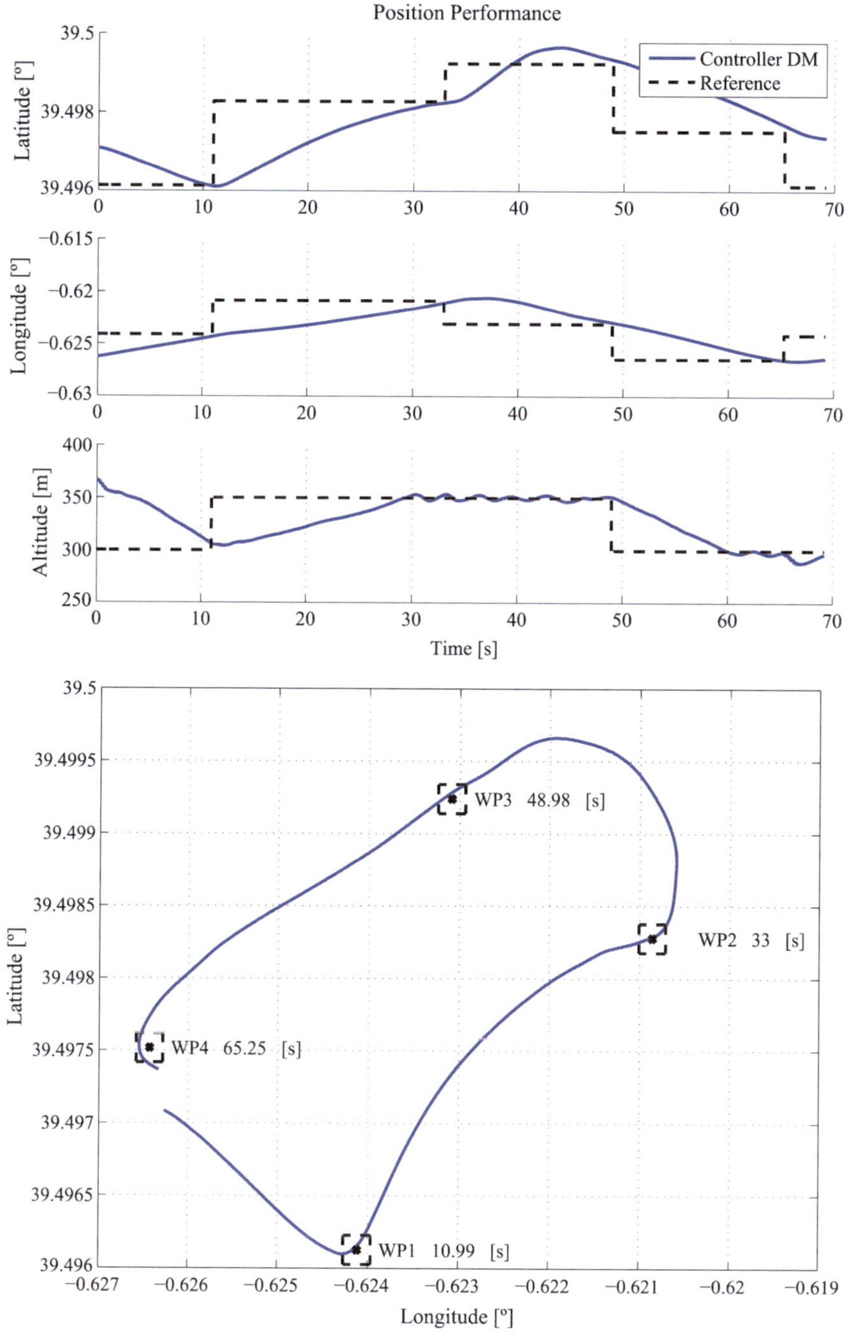

Fig. 12.8 Flight task performance of the selected controller $\boldsymbol{\theta}_{DM}$

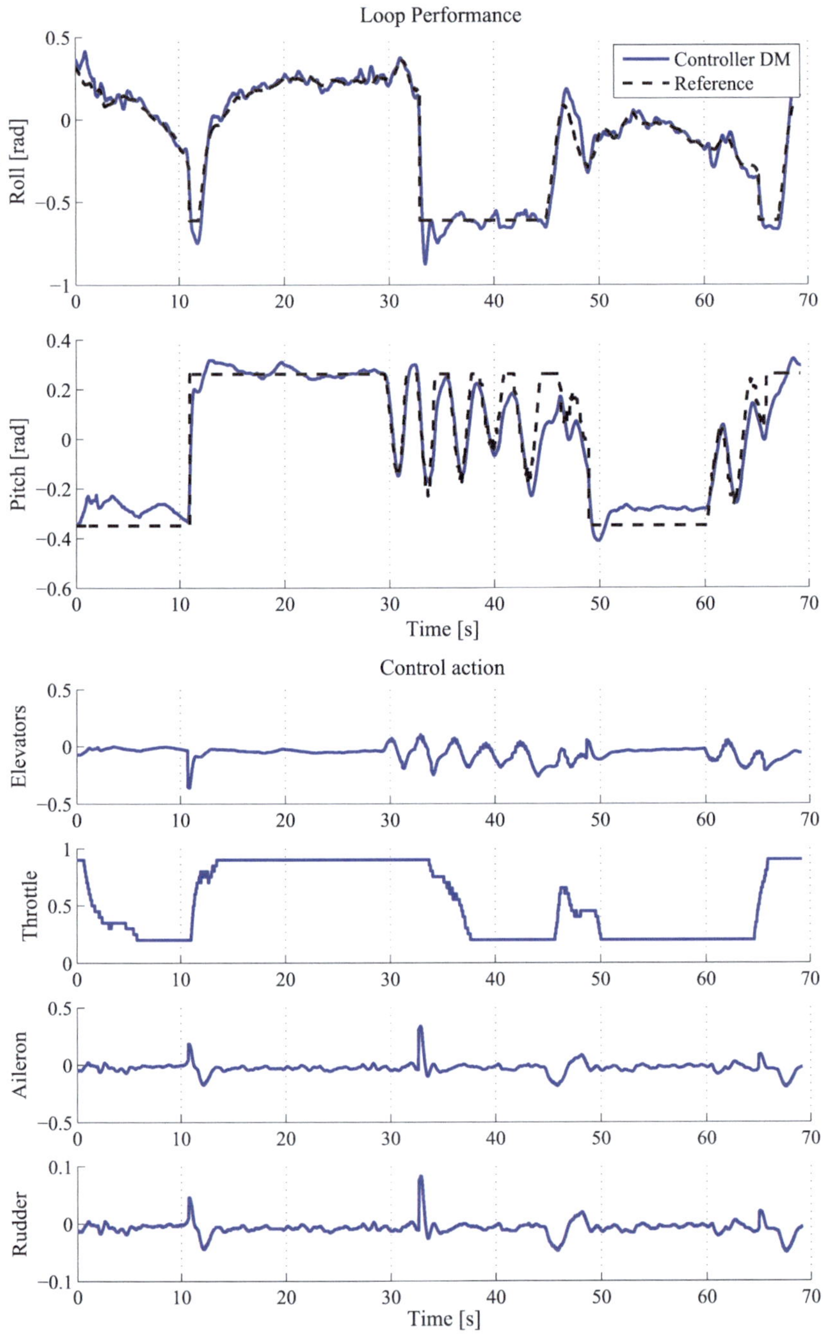

Fig. 12.9 Control loop performance of the selected controller $\boldsymbol{\theta}_{DM}$

12.5 Conclusions

In this chapter, a total of five PI controllers were tuned, in order to adjust the FCS of an autonomous aircraft. It was required to adjust a cascade control loop for altitude, a cascade control loop for heading and a simple control loop for velocity. A MOP with seven design objectives was stated, minding time performance and total variation of control action. As a result, a pertinent and compact Pareto Front was approximated. After an analysis, a controller was selected, implemented and validated in a real flight test.

References

1. CSS (2012) Unmanned aerial vehicle. Special issue. IEEE Control Syst Mag 32(5)
2. Du J, Zhang Y, Lü T (2008) Unmanned helicopter flight controller design by use of model predictive control. WSEAS Trans Syst 7(2):81–87
3. Fregene K (2012) Unmanned aerial vehicles and control: lockheed martin advanced technology laboratories. IEEE Control Syst 32(5):32–34
4. Kadmiry B, Driankov D (2004) A fuzzy flight controller combining linguistic and model-based fuzzy control. Fuzzy Sets Syst 146(3):313–347
5. Klein V, Morelli EA (2006) Aircraft system identification: theory and practice. American Institute of Aeronautics and Astronautics Reston, Va, USA
6. Pounds PE, Bersak DR, Dollar AM (2012) Stability of small-scale uav helicopters and quadrotors with added payload mass under PID control. Auton Robots 33(1–2):129–142
7. Reynoso-Meza G, Sanchis J, Blasco X, Freire RZ (2016) Evolutionary multi-objective optimisation with preferences for multivariable PI controller tuning. Expert Syst Appl 51:120–133
8. Song P, Qi G, Li K (2009) The flight control system based on multivariable PID neural network for small-scale unmanned helicopter. In: International conference on information technology and computer science, 2009. ITCS 2009, vol 1, IEEE, pp 538–541
9. Velasco J, Garcia-Nieto S, Simarro R, Sanchis J (2015) Control strategies for unmanned aerial vehicles under parametric uncertainty and disturbances: a comparative study. IFAC-PapersOnLine 48(9), 1–6. 1st IFAC workshop on advanced control and navigation for autonomous aerospace vehicles ACNAAV15Seville, Spain 10–12 June 2015
10. Velasco Carrau J, Garcia-Nieto S (2014) Unmanned aerial vehicles model identification using multi-objective optimization techniques. In: World Congress (2014), vol 19, pp 8837–8842
11. Velasco-Carrau J, García-Nieto S, Salcedo J, Bishop R (2015) Multi-objective optimization for wind estimation and aircraft model identification. J Guid Control Dyn 1–18
12. Wang J, Hovakimyan N, Cao C (2010) Verifiable adaptive flight control: unmanned combat aerial vehicle and aerial refueling. J Guid Control Dyn 33(1):75–87
13. Wargo CA, Church GC, Glaneueski J, Strout M (2014) Unmanned aircraft systems (uas) research and future analysis. In: IEEE aerospace conference, 2014. IEEE, pp 1–16
14. Zarei J, Montazeri A, Motlagh MRJ, Poshtan J (2007) Design and comparison of lqg/ltr and h-inf controllers for a vstol flight control system. J. Franklin Inst. 344(5):577–594

MIX
Papier aus verantwortungsvollen Quellen
Paper from responsible sources
FSC® C105338

If you have any concerns about our products,
you can contact us on
ProductSafety@springernature.com

In case Publisher is established outside the EU,
the EU authorized representative is:
Springer Nature Customer Service Center GmbH
Europaplatz 3, 69115 Heidelberg, Germany

Printed by Libri Plureos GmbH
in Hamburg, Germany